THE GREEN POWER OF SOCIALISM

History for a Sustainable Future
Michael Egan, series editor

Derek Wall, *The Commons in History: Culture, Conflict, and Ecology*

Frank Uekötter, *The Greenest Nation? A New History of German Environmentalism*

Brett Bennett, *Plantations and Protected Areas: A Global History of Forest Management*

Diana K. Davis, *The Arid Lands: History, Power, Knowledge*

Dolly Jørgensen, *Longing and Belonging: Recovering Lost Species in the Modern Age*

François Jarrige and Thomas Le Roux, *The Contamination of the Earth: A History of Pollutions in the Industrial Age*

Tirthankar Roy, *Monsoon Economies: India's History in a Changing Climate*

Elena Kochetkova, *The Green Power of Socialism: Wood, Forest, and the Making of Soviet Industrially Embedded Ecology*

THE GREEN POWER OF SOCIALISM

WOOD, FOREST, AND THE MAKING OF SOVIET INDUSTRIALLY EMBEDDED ECOLOGY

ELENA KOCHETKOVA

THE MIT PRESS   CAMBRIDGE, MASSACHUSETTS   LONDON, ENGLAND

© 2024 Massachusetts Institute of Technology

This work is subject to a Creative Commons CC-BY-NC-ND license.

This license applies only to the work in full and not to any components included with permission. Subject to such license, all rights are reserved. No part of this book may be used to train artificial intelligence systems without permission in writing from the MIT Press.

The MIT Press would like to thank the anonymous peer reviewers who provided comments on drafts of this book. The generous work of academic experts is essential for establishing the authority and quality of our publications. We acknowledge with gratitude the contributions of these otherwise uncredited readers.

This book was set in ITC Stone and Avenir by New Best-set Typesetters Ltd. Printed and bound in the United States of America.

Library of Congress Cataloging-in-Publication Data is available.

ISBN: 978-0-262-54745-1

10 9 8 7 6 5 4 3 2 1

To Saint Petersburg

CONTENTS

ACKNOWLEDGMENTS ix
INTRODUCTION xiii

1 ALARM OVER THE FOREST 1

2 THE INDUSTRY EXPANDS INTO SIBERIA AND THE FAR EAST 29

3 THE QUEST FOR A NO-WASTE ECONOMY 69

4 REED BECOMES WOOD VALUE 103

5 THE PRESTIGE OF MODERN TECHNOLOGIES 131

EPILOGUE: THE CONTINUITY AND DISRUPTION OF GREEN INDUSTRY IN THE (POST)-SOVIET ERA 157
NOTES 173
INDEX 205

ACKNOWLEDGMENTS

Dreams come true when you work hard. But working hard is possible only when you are supported by your social environment. Many of my colleagues and friends have contributed their intellectual and material resources to this piece of work.

As a historian, I am accustomed to looking back for origins. When thinking about the starting point of this book, I recall the year 2011, when Julia Lajus, Sari Autio-Sarasmo, Hanna Ruutu, and I discussed how forests in the late Soviet Union could be an interesting theme to study. In the years that followed, I began a long journey through the history of technologies, socialism, and the planned economy. I devoted most of my professional career to the history of technology transfer from the West to the Soviet pulp and papermaking industry. Over the last seven years, I have examined the interactions between state socialism and natural resources to write this book as a contribution to envirotech history of socialism.

My endless gratitude goes to Julia Lajus, Sari Autio-Sarasmo, Sakari Heikkinen, Andy Bruno, Laurent Coumel, Alexey Golubev, Alexandra Bekasova, Vasily Borovoy, Pavel Pokidko,

Aleksey Popov, Galina Orlova, Alexandra Balachenkova, Anastasiya Fedotova, Ari Sirén, Pentti Turunen, Evgeniya Veleva, and Elena Guseva along with the employees of the Segezha pulp and papermaking plant for their kind support in accessing materials and facilities as well as commenting on earlier versions of the monograph. I am also grateful to my colleague and friend Kirill Chunikhin, who once initiated a private discussion club on drafted books and book proposals. I was happy to dialogue about some chapters with my colleagues and friends Tatiana Borisova, Xenia Cherkaev, Zinaida Vasylieva, Nikolay Ssorin-Chaikov, and Roman Gilmintinov. I am especially appreciative of Xenia for tips on the prepublication process. Big thanks too go to Joel Pearson for his enormous effort in proofreading the earlier manuscripts of this book. And last but not least, I am thankful to Darina and Denis from Regensburg whose words "We will be waiting for a copy of your book" have kept me positive and inspired.

A few institutions have played a crucial role in this project. The Department of History and the Laboratory for Environmental and Technological History of HSE University at Saint Petersburg have long served as an inspiring environment for my work. The Leibniz Institute for East and Southeast European Studies (IOS) in Regensburg and particularly Guido Hausmann gave me a working space and encouragement to complete the book at the final stage. And my current institution—the Department of Archaeology, History, Cultural Studies, and Religion at the University of Bergen—gave me professional support to make my final manuscript a publishable book.

I am also grateful to the many professional communities and networks from which I learned a lot about global environmental, economic, and technological history. I especially appreciate the experience I had with the European Society for Environmental History (ESEH), where I served as a secretary between 2019 and 2021, Environmental History Now (EHN), and the Tensions of Europe Network (ToE), where I have been a member of the managing committee for several years. I met many great people involved there who helped me find my path in the field of environmental and technological studies.

My family has been a great supporter of my research. I am particularly grateful to Sasha, who was patient enough to listen to my long monologues about Soviet forests and technologies. And I offer thanks to Taisiya and Claudia, who kept me tuned in all the time.

Being sincerely grateful to all these and other colleagues and friends whom I forgot to mention, I dedicate this book to Saint Petersburg, the city where I had worked for many years and whose strong character and beauty will always be part of my heart.

INTRODUCTION

Our motherland is rich with nature.
—"Ne tol'ko rubit', no i vosstanavlivat'," *Master lesa* (July 1963)

Russians have traditionally considered their forests as inexhaustible.
—Brenton M. Barr and Kathleen Braden, *The Disappearing Russian Forest* (1988)

WOOD USE, PROFESSIONAL DREAMSCAPES, AND SOVIET PATH TO INDUSTRIAL ECOLOGY

In the Soviet Union, as in Russia today, it was typical to refer to forests as an empowering resource; huge forest coverage across the country made this gigantic polity extending across Eurasia "a green power" (*lesnaya derzhava*), a holder of natural "treasure" and "abundance." Forests had always surrounded people living there, and supported a widely held image of national power, political might, and cultural prosperity. Indeed, the USSR was one of the most forested countries in the world, possessing a vast array of different tree species—calculated at over 570 in the early 1980s.[1] Most of these forest riches were located in the eastern part of the country—Siberia and the Far East. According to some calculations, about 80 percent of the Soviet Union's forests grew in these regions. As a result, they were often described by Soviet commentators as both reserves of enormous green gold and "the pot of nature."[2] Present-day Russian state agencies similarly estimate that national wood resources equate to over a quarter of the world's supply and attribute a positive role to Russian forests in global climate conditions.[3] The Soviet-grounded image of a green sea of taiga has become anchored in popular sentiment, contributing to a sense of national pride.

Beyond imperial forest poetics lies their crucial economic service. Together with coal, oil, and gas, forests have played an extraordinary role in many of the world's economies. They provided construction and fuel material for industrialization, facilitating the growth of industrial enterprises, transport infrastructures, and housing. As such, wood was regarded as part of the foundation of modern societies of the last century. If before 1914 industries were capable of producing twenty-five hundred types of wood-based products, after 1945 they were capable of manufacturing up to twenty thousand. During the Second World War, wood offered a substitute for many of the materials required for producing certain key components in ships and aircrafts, and especially those made of metals—scarce resources in those troubled times. After the Second World War, wood continued to function as the framework on which grand scientific and technological achievements were premised. It supplied the material infrastructure for experimentation and discovery. One particular example can be seen in an "electric paper" invention, developed to record an image of the other side of the moon and photograph telegraphic messages from the atom icebreakers.[4] Bolstering the nation's pride in making modern technological material from a natural resource, the invention served to amplify the glorious achievements of the USSR in outer space while proving the military significance of wood production in modern times. In the age of high technological discoveries, the military and strategic application of wood was crucial. Wood also offered an alternative to modern polymers, and unlike oil, was a renewable and sustainable material. Cellulose, the material produced through the industrial cooking of wood, was used for manufacturing strategic goods such as gunpowder and military rubber, and enabled the essential material infrastructures for the war technologies at the center of the militarized economy of the Soviet Union.

The quest for modernity during the Cold War and the related growth in demand for mass consumer goods also made wood an important material. At this time, technology's role in society expanded due to its much wider applicability for consumption. New advances in wood processing, chemistry, and technology provided vast opportunities for transforming natural wood into synthetic materials and goods, surpassing the more technologically primitive use of wood for fuel, shipbuilding, and house construction.[5] Wood supplied the material for satisfying growing levels of consumption, providing people and their homes with numerous packages, plastics, cheaper furniture, and other commodities that had initially only featured in the United States and Western Europe. In the USSR, wood acquired a peculiar meaning in state-led attempts to develop a consumer society, starting especially with the rise of Nikita Khrushchev. As first secretary, Khrushchev placed particular emphasis on intensifying consumer manufacturing as a political project.[6]

The mass rollout of *khrushchevkas*—the new individual apartments pioneered by his government—demanded new material objects for individual consumers. And as a result, the state searched ever more urgently for cheaper furniture and construction materials to combat growing shortages in the Soviet planned economy.

This book considers the relationship between forests and industry under state socialism. It reconsiders what is known about the state socialist experience with nature, and shows the entanglement of the environment and economy. The USSR was a country underpinned simultaneously by a strong drive for the extensive exploitation of its abundant natural resources and desire for technological modernity. It logged and exported huge volumes of round timber to acquire the currency needed for purchasing industrial equipment and machinery. At the same time, the Soviet leadership saw strategic importance in developing sophisticated technologies to manufacture various valuable products from wood, this most versatile of materials in the context of the Cold War. This revealed the gap between the possibilities of the extractive economy and technological drive to maintain the lead in wood-based production, as captured in the well-known Soviet political mantra, "Catching up and surpassing the West." Declared in 1961, the goal of reaching Communism in twenty years was articulated as the main aspiration of Soviet industrial development and the scientific-technical revolution that became a key concept in the industrial discourse of state socialism.[7] To a large extent, this aim was premised on the material abundance and increased living standards of Soviet society. The notion of satisfying the consumer needs of modern citizens was part of the national political agenda in later decades, especially during Mikhail Gorbachev's perestroika, and was connected with the imperative to manufacture consumer goods from various natural materials, including wood.

The Green Power of Socialism stresses the activities of specialists working in forestry. It employs the broad term *specialists* to denote the industrial scientists, engineers, and wood-harvesting managers who worked at the harvesting and industrial enterprises, along with those employed at research institutions and administrative organizations related to the forestry industry. In the twentieth century, specialists, including technocrats and experts, played important political, economic, and environmental roles in both capitalist and socialist technocratic regimes. They participated in massive projects of nation building, contributing, for instance, to the technopolitics of attempts to build new, modern societies in places like Egypt and Francisco Franco's Spain.[8] In the Soviet Union, as in other countries over the course of the twentieth century, these specialists gained crucial power as technocratic voices, occupying a peculiar place in decision-making related to industrialization at various levels—in departments of the central level, industrial institutes, and individual enterprises. While most famously they formed the cornerstone of the Stalinist industrialization of the 1930s, they continued to play a significant role in industry building and economic activity long thereafter.[9]

Specialists exercised significant influence in advocating for the modern technological uses of wood and wood products, insisting that in the age of technological progress, "paper and cellulose, like coal and ore, are extremely important for our country."[10] They saw paper and wood-based products as crucial ingredients for technological and social progress—an idea that was primarily connected to modern consumption. While wood had served as a critical material in society and the economy for centuries, after the Second World War it became increasingly viewed as a modern material that could be put to much wider technological applications. While from the industrial and consumerist perspective, oil was undoubtedly a modern material, when converted from traditional to modern uses, wood came to be seen in similar terms. Thus as elsewhere following the war, modern science and technology rendered wood a liminal substance, transforming it from a *traditional* construction and fuel material to a *modern* raw material for the industrial manufacture of numerous consumer and military goods ranging from cardboard packages to the temperature-resistant cellulose used for making rubber for military aviation. The use of modern materials was seen to denote the pivotal change in the structures of production, consumption, and everyday life that underpinned notions of modernity.[11] Explaining the wide applicability of wood-based materials that became possible after the war, specialists responded to rapidly shifting models of consumption of natural resources.

These specialists were driven not only by real, empirically informed growth in wood demand but the anticipation of massive projected spikes in demand for wood too. With the beginning of the Cold War, many argued that the demand for wood would not decrease, despite the discovery of oil and progress made in the use of chemicals. Instead they predicted quite the opposite: technological advances would massively *increase* demand for wood due to the diverse possibilities of wood production, premised on the material's ability to change. Rapid Soviet technological progress in outer space engineering, atomic energy, physics, and medicine led many to believe that the forestry industry could also make a breakthrough and serve as a major provider of modern materials. Given the high military and consumer importance of wood, the second half of the twentieth century was punctuated by experiments to create modern technologies through the most efficient use of wood as an industrial resource. Like their Western counterparts, Soviet specialists saw sophisticated technology as a black box through which a raw material could be transformed into a ready product; through the operation of technological processes at socialist enterprises, the commodities required for a modernizing economy could be produced.[12]

While technology seemed to be the driving force behind the more intensive consumption of wood, the question of wood availability for developing this large-scale production became a key area of concern, however. Observing the rapid growth in wood consumption and expecting even more intensive demand, many specialists grew anxious about the sustainability of the Soviet Union's resource base in facilitating the technological age. This book shows that some of the ideas expressed by specialists about the environmental impacts of the Soviet exploitation of wood resources laid the ground for more ecologically sensitive production in the future, even as their advocacy dovetailed with industrial interests.

THE INDUSTRIAL DIMENSION OF SOCIALIST ECOLOGY

Wood was a natural resource that had greatly influenced the form that nature-economy relations had taken in the past.

The use of forests in extractive economies surfaced a key tension, though: to derive full economic benefit from this abundance required sophisticated technologies that had always been in acute shortage in the resource-dependent economy. The Soviet Union harvested significant portions of its forests and was one of the world's four principal exporters of round timber. Yet it made only a small contribution to global trade in terms of highly processed wood-based products.[13] Like other extractive economies, state socialism relied on wood as much as it depended on ore, coal, oil, and gas, among other natural "gifts." But wood exemplified the peculiar way in which the Soviet state and experts dealt with nature. It revealed the tension between the high military and civilian consumer demand for wood, on the one hand, and the problem of wood harvesting and processing, on the other, bringing to the fore one of the biggest challenges of the Soviet planned economy: *the prospect of the future scarcity of rich natural resources*. A scarcity of wood, lack of modern technology, and dearth of efficient forest management systems produced a new discourse of professional alarmism over the future of socialist nature and industry. Some forestry specialists argued that the apparent abundance of wood was rendered illusory if one looked at forests as they did, through an industrial lens. Rapid economic growth and the rising demand for wood, combined with what they conceived as inefficient harvesting and wood-processing practices, would lead, they warned, to the devastation of forest resources.

Over the course of the twentieth century and beyond, mass deforestation proceeded in virtually all corners of the globe at an alarming rate. The Amazon rain forest, for instance, has been subject to heavy devastation over the past fifty years, while some European nations, such as Denmark, depleted their own wood stocks much earlier. Soviet specialists rarely referred to the experience of other countries, but emphasized homegrown destruction and wasting practices. They warned not only against clear-cuts but also practices that produced vast amounts of wood waste—and especially those that left behind harvest waste in forests themselves. They criticized the rapid devastation of forests of the northwest and blamed it on *irrationality*, a term widely used in Soviet industrial parlance to signify economic loss. Indeed, the European part of the country had traditionally been heavily exploited, leading specialists to voice increasing concern about the possibility of the scarcity of industrial forests in the near future—particularly in light of the rapidly growing consumer demand, which was only expected to accelerate further. They also criticized the unequal geographic distribution of industrial operations, emphasizing what they referred to as the "weak" exploitation of forests in the eastern regions of the USSR, which remained largely industrially unspoiled. Imagining the forests of Siberia and the Far East of the Soviet Union as huge, untapped green riches, they insisted on the need to rationally industrialize them to halt depletion in the old industrial region of the northwest.

More specifically, some professional voices stressed that the vast promise that wood held in the making of goods could not be realized if the wasting practices of wood harvesting and manufacturing continued unabated. This densely forested empire was, they argued, putting at risk its share of the wood resources so crucial for facilitating modern consumer production. While for forestry specialists, nature offered important conditions for capitalizing on raw materials, they also emphasized the environmental limitations that came with the use of wood in economic production and consumption, underscoring the need for careful treatment and economic calculation. This shared concern connected forestry professionals working across the country, and led to a change in the model they used to describe the interaction of industrial goals with nature and the resources it provided. Specialists working in industry, the cornerstone of modern society, imagined and experienced nature as a key factor for technological development. They not only worked to realize the modern industrial and consumer society but acquired an important role in rethinking the relations between forests and industry as well. Reacting to the tension between the technological drive to develop a modern industry, on the one hand, and the extractive economy, on the other, some specialists highlighted the contradiction between rapidly disappearing wood stocks and the rising economic demand for consumer and military production. Some enterprises indeed suffered from a lack of wood, as they were located in deforested areas. This professional view challenged the public image of forest abundance and warned about the prospect of wood scarcity. Structured around the solutions that specialists proposed to the prospect of wood shortages, this book explores how specialists tried to *reconcile* nature, technology, and industrial production, seeking to manufacture modern products while preventing the depletion of industrial forests.

This analysis invites the reader to reevaluate interpretations of the relationships between the socialist state, industry, society, and nature. Scholars and the public have often previously explained these in one of two polemic forms:

ecocidal and environmental. One of the first substantial works on socialist forestry by scholars Brenton Barr and Kathleen Braden, for instance, took the first position, arguing that Soviet exploitation of forests was wholly destructive. They showed that forests offered the state timber for export, through which it could accrue the currency needed for purchasing costly Western machinery.[14] For Barr, Braden, and other scholars, the Soviet approach to nature resulted in "a vast, toxic rust belt of chemical, metallurgical and nuclear factories and extractive industries spewed smoke, acid and poison into the air, water and land over decades of Soviet power."[15] Soviet practices echoed the experiences of other socialist countries like the German Democratic Republic, where political priorities led to "inevitable forest collapse."[16]

This scholarly view has been forced to reckon with emerging evidence that suggests that dictatorships can be more environmentally friendly than previously thought, decoupling the association of totalitarianism with environmental destruction.[17] In the case of the Soviet Union, some have come to describe socialist development as having been influenced by forms of environmentalism, either as a type of Soviet intellectual activism or state policy, and part of tensions around particular natural assets arising from industrial construction. Important and widely known events, such as the "storm over Baikal" in which protests erupted over industrial construction on the shore of this unique lake in the 1960s or the Chernobyl catastrophe in 1986, were triggers for scholarly and mass media discontent with Soviet technological politics.[18] They provoked late environmentalist attitudes within Soviet society, just as earlier criticism of agricultural chemicals caused anxiety in the United States.

The relation of authoritarian regimes to nature was more than one-sided, as the model of ecocide previously suggested. Some research is indeed more positive about the socialist (and more broadly, authoritarian) experience around the use of nature, even though one stresses the general technological backwardness of industry and thus wasteful forms of natural resource exploitation.[19] The utilitarian view of nature espoused by Soviet specialists as primarily a source of economic value came to overlap with a drive for developing a more holistic approach to the natural world.[20] This lent views of nature a sense of hybridity that emphasized the nexus between the natural world and technological infrastructures. The industrial ecosystem came to be understood as "a transcendental hybrid" of natural and artificial systems when engines included nature in the calculus of industrial production.[21] Now that the ecocide narrative that posits the economy swallowing nature has been considerably challenged, historians have been stressing the combination of exploitation and protection that appear in Soviet discourse.

This book offers a more robust interpretation of economy-nature relations under state socialism. It suggests that among scholarly analyses, the role of specialists working in industry deserves more discussion in the context of complex relations with nature. Existing bodies of work often depict specialists as technocrats who played different roles: decision-makers in their own right, victims of political decisions, or agents of global communication between the East and West.[22] Their place in Soviet economic policy and global agendas was enormous. But this book proposes examining socialist specialists in another light, focusing on their relation with natural resources. Furthermore, unlike much previous scholarship that concentrates on forest protection (in particular on *zapovedniki*), environmental activism, and forestry legislation, this book discusses strands of socialist environmentalism by looking deep inside industry—the heart of the Soviet project. It examines how industrialists saw the past and future of forest exploitation in light of industry-nature relations.

These relations, this book demonstrates, were constituted by the combination of economic interest and ecology, highlighting the controversy of industry-nature interactions. Industrial specialists working under state socialism explained the trends prevalent in forestry and wood use as rapidly moving toward a *wood crisis*, and this stimulated an intensive search for new raw material resources from the early 1950s on. From this perspective, the Soviet project saw itself as not only a heroic movement toward Communism in which humans were victorious over nature but also filled with both economic and environmental risks. *Crisis*, which as historian Rosalind Williams has recently argued remains a hazardous term, proved a powerful category and found advocates in the Soviet Union.[23] Forestry specialists, moved by concern over the industrial future of the extractive economy, grew alarmed about the Soviet Union's forest stocks. The sense of crisis derived from the expectation of wood scarcity expressed by specialists in the context of growing economic demand, on the one hand, and intensive but wasteful woodcutting, on the other. The desire to compromise more for industrial purposes led to a less devastating orientation toward nature among specialists, or what this book calls the *industrially embedded ecology* of Soviet state socialism. This shows that specialists who worked in industry were not a priori killers of nature nor were they its fervent protectors. Instead they evinced a complex vision of nature and its resources, framed by imperatives of economic and industrial growth. Wood, as an economic substance with a strong technological effect and wide applicability—both natural and technological—entangled industrial and environmental issues. Specialists insisted that forests should be treated carefully (or in their language, "rationally") in order to stop wastage and maintain a sustainable base of raw materials for the expected increase in production. Soviet industrial ecology represented a dimension of the Soviet environmentalism growing within the industry and stemmed from the envisaged increase of economic consumption that would, if practices of wood harvesting and processing remained unchanged, lead to a wood crisis. This shows how a form of industrial ecology was born from a productivist view of nature and overindustrialization.

Importantly, professional conceptions of industry and forestry were not informed by purely technical conceptions but quite often involved an emotional response on the part of experts sheathed in a studied professionalism. This can be seen in the invocations they made around the transformations of forests and wood, and waste and annual plants, in their descriptions of past and future wood stocks, state policy, and industrial experiments. Their actions were significantly colored by their professional imagination and expectations of the future as well as the comparisons they made between Soviet forest practices and those of other countries.

This included *alarmism*, the term this analysis uses to describe concern about—and even fear over—the future of industrially useful forests. Alarm about the prospect of shrinking natural resources moved many to advance claims and make decisions to improve wood harvesting and processing. More broadly, it changed their relation with nature. It produced enthusiasm for and a belief in the industrial opportunities created by resource colonization and technological experiment. The great hopes in turn evoked disappointment with the frustrating results achieved in Siberia and the Far East. The forestry industry, as this book shows, was not a purely technical phenomenon but rather a space of technocultural emotional responses to deforestation. Emotional responses and perceptions of wood availability among specialists, who drew on scientific and industrial investigations, produced industrial processes penetrated with expectations, fears, and hopes that pushed forward certain economic and political decisions. This was important for how specialists participated in the setting of agendas about natural resources along with their availability and use. Specialists, however, were the generators of expert knowledge, explaining the parameters of crisis and proposing solutions from their own perspectives.[24] The chapters that follow examine industry as a space of technologies, critical problems, and decision-making deriving from responses such as alarm, fear, hope, and expectation.

The relationship between specialists and nature in the last century correlated with what political scientist James C. Scott has called "high modernism": interests and faith conducted through state action.[25] In state-led and sometimes market economies, the state provided investments and labor for advancing to the unspoiled lands and building experimental enterprises. Indeed, solutions to the wood crisis required huge investments and centralized action—to build technological and social infrastructures in difficult-to-access forests, construct new industrial factories and research stations for experimenting with alternative raw materials, and modernize available forms of technology. Many forestry industry specialists combined expertise and administrative responsibilities working in state institutions and thus were state officials themselves. Looking at professionals and industry, this book also explains the development of state politics and decision-making in the planned economy.

Considering the technoenvironmental dreamscapes of specialists under state socialism illustrates how they measured economic geographies in terms of the availability of industrial forest resources as well as the technological possibilities to render them both more sustainable and productive. They dealt with various material substances, both dead and alive: forests as living species; timber, wood, and ready-made consumer products; and alternative resources such as harvest, sawmill, paper waste, and annual plants applicable in the forestry industry. All of these materials were the subject of physical and symbolic transformations that made up a modern variant of production. Hence the changed paradigm of nature-industry relations, and new technologies to manufacture wood pulp and cellulose, made wood waste and annual plants (such as reeds) valuable economic materials and stimulated numerous industrial experiments. Consequently, they provoked the enthusiasm of many specialists about a new age of making wood-based products without the intensive felling of trees. Importantly, they did not address wood, alternative raw materials, and manufactured goods as actors but instead as material resources for achieving progress and modernity to overcome an outdated state—goals that mattered a great deal in the calculus of the

Cold War. Forests were likely a source of valuable materials in their dreamscapes, but by the end of the epoch, specialists increasingly spoke about forests as living organisms and actors within complicated ecosystems that had not simply national but indeed global impacts.

EXPLOITING FORESTS, SAVING FORESTS

Three solutions, proposed by Soviet specialists in the 1950s, grew out of economic interest and concern about the disappearing industrial forests. These solutions can be summarized as follows: the imperial and extensive advancement to the eastern lands of the country; the development of no-waste technological production; and technological improvements in wood harvesting and processing. Advanced by different groups of specialists working across the country, these notions nonetheless represented overlapping responses to wood scarcity. Socialist specialists in this sense deployed their power to find critical solutions and develop environmentally compatible technologies for keeping industrial growth sustainable. In professional dreamscapes, alarmism about the future of wood resources provoked a search for ways to preserve natural abundance.

The imperial solution harnessed the ambition to exploit new—meaning unexplored—forests in Siberia and the Far East, facilitated by large-scale railroad construction along with oil and gas excavation. This unfolded in the course of the postwar colonial-industrial turn toward the eastern Soviet lands.[26] The late 1950s and 1960s represented the peak of Soviet discovery of carbon raw materials and heralded the construction of the famous Baikal-Amur Mainline

(BAM). It supported yet another episode of the colonization of the eastern territories, now much more technologically equipped and significantly more focused on industrial construction than it was in the nineteenth century when the government of the Russian Empire expressed a colonizing interest in its eastern regions. Thus in the mid- to late twentieth century, the Soviet Union proved itself an inland colonial empire moved by economic ambition. Forestry specialists, however, urged rational forest exploitation in the east to start a new page in wood harvesting and industrial production, advising the state not to repeat the negative results that had followed the exploitation of European and particularly northwestern forests, where the industrial wood stocks had been depleted. Rationality implied economic efficiency, and was connected to the so-called complex use of natural resources and production, underpinned by the concept of no-waste industrial manufacturing. This approach was partly implemented under the slogan "enterprises of the future," as Soviet propaganda and specialists themselves put it, or forest-industrial complexes (LPKs), many of which were indeed established in the Far East. Broadly, the availability of eastern riches and the state's turn toward them increased alarmist views, which referred to the fate of the overexploited northwestern forests. The rational use of natural resources referred to modernity, and was embodied in the struggle against technological backwardness and the ineffective use of resources extracted from nature.

The second set of solutions was underpinned by the discourse of complexity in Soviet forestry rooted at latest in the 1930s. From this time on, specialists began to advocate for the importance of using every possible element of natural resources and not leaving behind anything as waste. Yet at that point, the industry suffered from a lack of appropriate technology. After the Second World War, threads of this approach were resurrected in the search for wood alternatives, including various wastes and annual plants, leading to significant changes in the consumption of resources and their economic sustainability. This alternative industrial production was primarily designed to foster a base of renewable industrial resources for the unforested southern regions, such as Ukraine, Kazakhstan, and the south of Russia, drawing on their stocks of reeds and other annual plants. Such approaches were, however, also employed in the old northwestern region and especially the new eastern forest regions. This tempered imperial ambition in the east, encouraging the more intensive use of waste and other available resources, the use of which would be less harmful to nature. But this situation was made more difficult by the fact that notions of the complex and rational use of resources were connected with imperial designs; it was in the course of discussions around the technological advancement into the new regions that specialists considered ways of using consumer (paper) and industrial waste in manufacturing processes at new enterprises. The Soviet authorities devised schemes of industrial construction that implied the building of a network of enterprises in both the densely forested east and sparsely forested south. This approach also aimed to solve the problem of supplying unforested regions, namely Ukraine, Kazakhstan, and the south of Russia, with renewable raw materials and ready manufactured goods in situ, thus decreasing transportation costs. Specialists conceptualized numerous alternative materials, ranging from industrial, wood, and consumer waste to annual plants as wood substitutes to cope with the growing demand for wood-based products. Moreover, "used natural resources" or "natural remnants" were now conceptualized as *modern* materials of industrial production, which held potential as cheaper, more flexible, and transformative industrial resources. Experiments with alternatives were designed for the regions that were unforested or where forests were rapidly disappearing, but they were proposed as an important condition for the colonial exploitation of new forests to prevent the transfer of old negative practices of wood use to the newly opened eastern lands.

Finally, the third set of solutions proposed by specialists to deal with the wood crisis centered on attempts to develop practices of more efficient wood harvesting and industrial production in the forestry industry. This included the attempt to improve technology and methods of work within the industry by enhancing mechanization, automation, and the quality of manufactured products. It stemmed from expanding technological possibilities that substituted human muscles for mechanisms, a more or less continuous process from at least the 1920s and 1930s onward. Stalinist industrialization emphasized the significance of the mechanization of heavy works in forests and at industrial enterprises, particularly those related to loading and transporting raw wood and ready-made products. Later Soviet decades saw a more sophisticated approach stressing that all operations could be governed by machines. Specialists believed that due to their technological accuracy, mechanisms could make the harvest and consumption of wood more precise. Through technological modernization and automation, many hoped that efficiency would increase and wastage would be minimized during processes of harvesting and industrial manufacturing. Together with the imperial and experimental solutions, the modernization solution implied huge investments in production in order to increase the productivity of wood supply and consumption.

Many of these approaches were not especially new. The notion of the sustainable use of forest resources dated back to nineteenth-century Europe, for instance, when foresters discussed the stable growth of trees. In addition, the concept of the complex use of natural resources—a notion this book will discuss in some detail—was popular in industrial discourse of the 1930s in the Soviet Union. Some projects were interrupted by the Second World War, though, while others could not be implemented due to a lack of proper technological infrastructures. In their conversations, late Soviet specialists often referred to the 1920s and 1930s, when their counterparts (and their younger selves) had worked on more efficient methods of using raw materials for the industry.[27] In this sense, many of their projects were not innovative but instead developmental. Yet there were differences between the post-1950s and earlier epochs in at least two respects. First, new technology made many ambitious projects possible after the war, giving the industry new sets of instruments for experimenting with wood. In the wake of growing consumerism, postwar technological advancement in forestry chemistry in particular helped render wood a more flexible material than before. The marriage of chemicals and wood could help satisfy demands for consumer products such as cloths, paper and cardboard packages, and cellulose-based soluble tablet shells, among many others. Technology also provided the possibility for the automation of harvesting and industrial operations, making them more effective and less damaging to the environment. Second, professional attitudes toward nature resonated with developing environmentalism as a reaction to the intensive consumption of natural resources. Even as it remained concentrated on economic growth, the late Soviet industrial approach to nature evolved from exclusively industrial concerns to one that comprehended forests as complicated natural organisms. Later decades of the Soviet Union therefore exemplified how notions around forests were being transformed in professional dreamscapes: from an endangered industrial resource, forests came to be increasingly understood as a natural organism under threat. This became especially obvious by the 1980s, when many specialists emphasized the liminality of forests, addressing them as both a complex source of wood and factor impacting the environment.

As the following chapters will show, the images of future resource availability and the search for solutions to an impending wood crisis led specialists to conceptualize forests not only as a hybrid industrial material produced by nature and consumed by people but as a *finite* industrial resource of nature too. The book builds the argument that with a utilitarian view of nature, and searching for the rational use of natural resources for the economy and society, Soviet industry developed industrially embedded ecology as a *by-product* of hyperindustrialization. Placing industry at its core, state socialism aimed to achieve not only sustainable industry but also sustainable forests to provide a continual supply of resources to industry. This approach derived from earlier periods and flourished after the Second World War. By the 1980s, while nature was still understood in its service to the economy, it was nonetheless to be treated with care. Taking economic growth as a nonnegotiable imperative, socialist industry was thus increasingly sensitive in marking this natural resource as a fragile industrial material whose exploitation required more careful approaches. Rethinking the past, present, and future of industrial forests and practices of wood consumption provided the impetus for reconsidering their economic function along with the shifting of focus to alternative resources, the more efficient use of wood, and attempts to stop the devastation of old forests. As such, forests were rethought not simply as industrial value or a sign of imperial might but as a *natural resource* of state socialism that was *at risk* as well. At the same time, while forests were increasingly recognized for their environmental value, waste and alternative resources accrued *industrial value* as substitutes for wood in industrial processes. Industrially embedded ecology thereby grew within the industry itself—a result of the industrial alarmism that suffused specialists in the sector who aimed to redirect the industry toward sustainable use. Hence Soviet specialists reconsidered the relation of industry and nature, seeking to render forests more productive resources by substituting natural wood with wood waste and annual plants in industrial production. Many specialists expected that further industrialization along the same old lines would perpetuate the wood resource crisis and so the search for alternatives was vital.

Out of the three alternatives proposed to solve the wood crisis—the imperial, experimental, and modernization approaches—none emerged as the singular winner. Instead, all demonstrated the desperate industrial search for saving forests from depletion. They contradicted the conventional view of Soviet forest abundance, undermining the more general view of enormous stocks of wood resources in the country. While in Sweden, for instance, cooperation between industry, the state, and ecologists was fairly fruitful in terms of making production more ecological, in the Soviet Union it remained a matter of professional discourse. In practice, many of initiatives, given their cost, depended on state planning and action. The planned economy was eager to develop new technologies for the more efficient use of forests, but was itself frequently an obstacle and reason for the failure of specialists' initiatives. It did not succeed in investing enough resources into the reorganization of forestry as many professionals expected.[28] On the discursive level, however, this story illustrates that state socialism was not exclusively a space of ecocide but instead characterized by a much more complicated set of relationships. This leads us to reconsider socialism from an environmental perspective—a task that bears significant relevance today.

With the demise of the Soviet planned economy, the discourse of industrially embedded ecology disappeared almost entirely, demonstrating a rupture between socialism and postsocialism. If the postwar Soviet era showed continuity with previous periods, reviving some earlier initiatives in approaches to wood, the post-Soviet period disavowed the environmental approaches developed during the Soviet period. In the early 1990s, the forestry industry was in a critical state due to a significant lack of funding and declining numbers of employees. Russian and foreign logging companies exported large quantities of wood with little consideration of reforestation, neglecting Soviet experience as a remnant of Communism, the relic of an odious past. Importantly, many issues voiced by Soviet specialists, such as nonwasteful wood harvesting and the use of wood waste, were largely forgotten even as the Russian economy has remained enormously extractive. Despite promises, private companies have neglected expensive and time-consuming projects—for recycling waste and experimenting with the use of alternative materials, for example—in favor of economic profit. Some aspects of logging have been undertaken by illegal companies that have contributed massively to levels of deforestation through voluminous exports of raw wood. Discussions around alternative resources and decreasing deforestation are currently unfolding in modern Russia and beyond. Yet few make reference to the Soviet past, seeking instead to find their own solutions. The fall of socialism thus created a rupture that saw forests—one of the major representatives of Russian nature—suffering significant losses. This book aims to facilitate a process of rediscovery, bridging this disruption by tracing processes of industrial "ecologization" in state socialism and its aftermath in postsocialist Russia.

1

ALARM OVER THE FOREST

Gifts of a green friend.
—Anatoliy Averbukh and Kseniya Bogushevskaya, *Chto delaet khimiya iz drevesiny* (1970)

INDUSTRIALISM, ALARMISM, AND FOREST GEOGRAPHY

"Our Motherland has a rich nature," proudly declared *Master lesa* (Forestry expert), a leading journal of the Soviet forestry industry.[1] This text, published on the front page of the summer issue of 1963, made passing mention of the need for reforestation. Yet it tapped into an entrenched and widespread national imagination that saw forests as endless riches that belonged to the whole society—"our forests." It stressed that forests were being put to the service of the country, and were, as ideologically proclaimed, the cornerstone of national material prosperity under state socialism. In fact, in this noncapitalist economy, *the state* was the main consumer of wood, using it for industrialization while *ideologically* framing it as the forest riches of the people.

Many Soviet commentators compared national forest stocks with the physical size of other countries, measuring them through a geopolitical lens. Thus a forestry engineer wrote in 1961 that the size of taiga forests was equal to that of the whole of England.[2] In 1964, another commentator argued that "there are so many forests on the territory of the Irkutsk region [in Siberia] the size of which is equal to the territory of Norway, Sweden, Finland, and Yugoslavia taken all together."[3] Quite often, publications replicated ritual images and referred to the size of capitalist countries to demonstrate the richness of Soviet nature in contrast to the scarcity of nature under capitalism. Publications particularly described the eastern regions of the USSR as a national treasure (*kladovaya*) and proclaimed that the "all-Union wellhead of wood" would be necessary to explore in the nearest future.[4] Geographic size therefore mattered a great deal in the framing of forests, and produced a strong belief in endless green covers and the vast economic possibilities they could open up. It triggered the image of forest abundance and inexhaustible industrial resources, kindling national pride in the extractive economy.

Specialists expected this resource abundance to offer numerous opportunities for industrial production, the material foundation of the modern society. From the beginning of the Soviet state socialist project established in Russia in 1917, forests were viewed through an industrial lens: they were to be used to their utmost potential in the aims of intensive industrialization.[5] In 1931, the government divided forests into industrial and nonindustrial or protected categories, prioritizing the economic function of forests, as the former category of forests was much more numerous than the latter. In 1943, a government decree further divided the state forests into three groups: group I forests were protected zones in which cutting was prohibited; group II denoted the sparse forest stocks of some parts of the southern and European regions of the country where partial cutting was allowed; and group III forests were the largest group of industrial forests and were to provide resources for economic needs.[6] In the decade after the Second World War, Soviet publications contended that the purpose of the forestry industry in using these resources was to satisfy the consumer needs of the country because, as one specialist put it, "forests must necessarily give the country an economic effect."[7] From his perspective, one shared by many industrialists, forests were troves of potentially valuable materials that had crucial economic meaning for the state and society, supplying both with the fuel for industrialization.[8]

The planned system tried to harvest as many natural resources as possible according to the logic of five-year economic plans. Wood harvesting was declared an urgent economic task. "Comrades, Give More Timber to the Country!" was a common slogan at that time.[9] This approach was rooted in conceptions of forests as offering important materials for facilitating production and revealed that industrial discourse in resource consumption was predominant. This view originated in earlier decades, revealing continuity between pre- and postwar periods in how the interplay between nature and industry was understood. For example, a 1930 poem by the children's writer Samuel Marshak titled "The Holiday of the Forest" (*Prazdnik lesa*) began with a question: "What do we plant when planting forests?" Answering this, Marshak listed a number of items of industrial construction: by planting a tree, we in fact plant masts and yards for ships in order to travel across the sea, wings to fly (airplane wings were largely made from plywood at that time), a table, and pencils; only later does Marshak emphasize the role of the forest as a home for animals and source of "morning freshness." This poem prioritized a consumerist perspective of the forest and conceptualized it as a source of industrial wood, implying that forests should be respected not primarily because they were living organisms of the earth, central to the ecosystem, but rather because they provided a great economic service.

This attitude persisted after the Second World War; in 1953, the professional journal on wood processing wrote that Soviet people "take all their pains to use enormous reserves of the soils of our regime as most intensively as possible."[10] And later, another professional journal wrote, "The country does need timber! Tomorrow it will be turned into furniture for new house dwellers, sawmill materials for builders, standard houses for countryside workers. Everyone must make most efforts and creativity, [and] high labor consciousness to complete the plan on wood harvesting."[11] Engineers attributed to the state a power and right to exploit forests; to take one example, Yakutia, a huge region in Siberia, "is proud of its diamond excavating industry created by the Soviet power. Its main riches include also forest resources."[12] For industrialists, this statement revealed a desire to be the first in economic achievement—aspirational rhetoric typical of the socialist project in general. In the context of the Cold War and East versus West competition for modernity that underpinned it, the regime saw wood as an important factor in beating the West. Both the state and specialists made frequent comparisons with the West (and the United States in particular), explaining how important it was to overcome the relative

Figure 1.1 Soviet matchbox label, "forest is our treasure," 1971. *Source:* Match Museum, http://match-museum.ru/catalog/320/2597.

backwardness of the Soviet forestry industry in order to provide a source of modern consumer production. Industrialists pointed to the availability of unexploited forests, especially in poorly investigated eastern parts of the country, and insisted on the significance of industrial advancement there. Modern technology was held to be the means of advancement, and played a crucial role in mediating between rising consumerist demand and the state's wood stocks.

Yet while the image of green abundance remained widespread, the economic expectation and evaluation of current *industrially* available wood stocks had a polarizing effect. Some specialists working in the forestry industry thought not only about industrial possibilities but also calculated the economic risks of intensive resource use. From the mid- to late 1950s, a new strand developed within industrial forest discourse that gave the green light to much more critical and less utopian thinking about the forest as an endless economic resource. At that stage, some time before specialists had seen the results of industrial advancement in Siberia and the Far Eastern regions (discussed in chapter 2), this alarmism mainly concerned the northwestern forests, which had traditionally been intensively exploited. Some specialists looked back to the prewar past of forestry in Russia and critically reappraised long-held ideas about forest abundance. They argued that the image of forests had to be reconsidered: the green covers of the Soviet Union could not so simply be understood as an industrial abundance since many forests were overexploited and cut down; others remained difficult to reach and exploit industrially. Noting the contradiction between imagination and rationality, they stressed that the image of forest abundance in the Soviet Union was a cultural myth devoid of rational industrial calculation. Hence as a 1962 book on forests insisted, "*Unwittingly*, we get an impression about the inexhaustibility of our forests. The attitude towards them is . . . a sort of something eternal, forever given, and abundant . . . but this is not really true."[13] They underscored that the northwest of the country had been overexploited while the green eastern regions had been underexploited and required further exploration. They estimated that eastern Siberian forests, for instance, made up three-fifths of all wood stocks. The growing consideration given to exploiting eastern forests triggered the recollection of recent experiences of rapid deforestation in the northwest, however. And, they asserted, it was crucial to turn to a new page in resource exploitation by developing new practices to stop ineffective harvesting. This stemmed not simply from their own desire, they maintained, but from objective necessity too, "dictated by the high cost of wood and depletion of fir raw wood" in particular.[14] Some complained that industrialists had previously cut the most valuable fir trees along highways, and that this practice must be stopped because it was backward and literally devastating.[15] The argument about preventing the wastage of wood became important for specialists who claimed that forestry politics had to change and the exploitation of forests had to become more effective.

This specific view was expressed by specialists from various institutions alarmed about the coming resource scarcity in the rapidly industrializing economy. In fact, the major problem that arose in dealing with Soviet forests as industrial resources lay in their uneven geographic spread and the historical background of wood harvesting. As in other countries (such as Sweden), the geographic distribution of Soviet forests differed significantly from region to region for both natural reasons and because of long-term historical cutting practices. Professional concern over the future of the industrial wood supply referred to three large regions in relation to their forestation and consumer demands. First, there were the northwestern forests that rose to the Urals (what I call "old forests"), which were historically used for intensive economic development. This part of the country was traditionally more economically developed and populated than the Far Eastern or Asian regions to the east of the Urals. By the 1950s, the share of the Russian Republic in wood harvesting was most crucial, making up to 90 percent of all the wood harvested in the USSR, with the northwestern region providing the largest share (about 25 percent). Second, there were the huge forest stocks of Siberia and the

Far East, which became the subject of large-scale intensive exploitation (what I call the "new forests"). These were not easily accessible, requiring, as one article put it, "[strong] will and desperate work effort" to start exploiting them.[16] These forests were often used as the primary evidence in arguments that described "the unexplored abundance" of Soviet natural resources. These, many believed, could be the savior of the rapidly disappearing northwestern forests.

The third broad geographic zone of professional concern was made up of the southern regions of the country, which were mainly unforested, but as in the northwestern region, fairly densely populated and required supplies of consumer products, including paper and cardboard. These three regions defined Soviet forest geographies: the northwest, characterized by technological overexploitation; the untapped green east, with its lack of harvesting infrastructures; and the sparse south, lacking in wood. Among these, the Ural region was also important as a traditional industrial region, but specialists did not refer to it frequently in their proposals presumably because its forests had lost most of their industrial potential by midcentury. This prompted many specialists to think about a coming crisis in the harvesting and supply of industrial wood, a material important for the building of modern society. The situation was complicated by the historical location of the Soviet Union's main forestry capacities: due to intensive construction in the age of industrialization, its main forestry enterprises were built in the northwestern parts; there were no large-scale industries in Siberia and the Far East before the Soviet leadership turned its attention to these lands.

The technological changes that opened up possibilities for specialists to manufacture diverse products from wood and spurred growing consumer demand for numerous products, such as food packages and "sanitary" (toilet) paper, changed specialists' view of these three economic regions after the war. This change was supported by rapid postwar economic growth in the Soviet economy, which only began slowing down in the mid-1960s. Great technological achievements created strong enthusiasm among specialists around the potential of exploiting natural resources more successfully. Forest alarmism implied that Soviet space was not as green as initially believed from an industrial perspective; against the backdrop of growing consumer demand for wood-based products, the amount of industrially appropriate forests was rapidly shrinking while the rest remained inaccessible. Fir trees were the main targets for exploitation and export, while larch trees were most prevalent in the country yet least utilized in Soviet industry. Typical of the Soviet habit of making prognoses and planning, all specialists expected large-scale growth in the consumption of paper- and pulp-based goods and even more rapid decreases in available wood, connecting the demand for wood with technological progress and the influence of evolving forms of knowledge on industrial manufacturing. For example, some expected that the production of chemical fibers would increase fourfold while the manufacture of plastic masses and synthetic materials would increase sixfold.[17] They envisaged a wood crisis stemming from this growth—a threat to sustainable forestry production that could lead to wood scarcity.

Alarmism triggered a search for solutions and alternative sources of raw materials for increasing wood consumption instead of intensive cutting in the northwest. As one specialist said in 1962, "The need for enlarging the production of paper makes us reconsider the possibility of using other sources of raw materials, in particular, in the regions where the lack of wood is evident but where other reserves are available too."[18] Alarmism over the lack of industrially useful wood was described variously in different regions of the country: in the northwest due to recognized overexploitation, in the east due to the inaccessibility of forests, and in the south because of the natural lack of forests.

Importantly, alarmism was born within industrialism. In other words, anxiety about the scarcity of the future resource base was triggered by an expectation of massive production growth. At the same time, this alarmist view was not a purely industrial phenomenon but also entangled with public criticism of the industrial exploitation of nature. Alarmism among employees of the forestry industry resonated with a more general (yet still nascent) public concern about the destiny of forests and their future. The 1953 book by writer Leonid Leonov titled *The Russian Forest* in particular, and its 1964 film adaptation, emphasized the environmental problems stemming from intensive woodcutting.[19] By contrast, the industrial forest alarmism along with the search for alternative methods to replace devastating harvest and wood-processing practices were connected explicitly to the need for sustainable yields of the resource, thereby still directed by the drive for economic profit. Specialists insisted that forest loss and the low productivity of wood were caused by technological backwardness and a lack of investment. In addition, uncleaned logging spots and the general wasting of forests was held to pose "a danger for the wide spreading of vermin in the forests" and "increase[d] the probability of forest fires."[20] Alarmism convinced many specialists that a future wood crisis was inevitable unless changes in wood harvesting and consumption were made.

EXPLAINING THE "WOOD CRISIS"

Alarmism among industrial experts suggested that demand was growing against the backdrop of low forest productivity in the USSR. Comparing Soviet forestry with the industries of other countries, some specialists admitted a gap between the availability of natural resources and technological possibilities. They recognized that the Soviet Union was lagging behind Western forested countries technologically and compared the Soviet performance with the Western world in terms of inequality. They measured their own backwardness through the lens of a crisis in wood supply because of wasting practices associated with harvesting, transporting, and storing wood. Specialists described wasting as the biggest evil in Soviet wood harvesting. Harvesters indeed left huge amounts of wood waste, like bark, branches, and roots, in forests. As such, some Soviet specialists portrayed old industrial forests as "a cemetery of the forest," referring to the fact that logging spots were not cleaned after wood harvesting and became forest waste.[21] They connected this practice to the "wrong culture" that had developed around forestry, and associated it with economic loss and low productivity. If the commentator Donald Bowles is to be believed, the daily productivity of Soviet wood-harvesting enterprises in 1956 was equivalent to just one-third of the productivity of similar US enterprises because of the low level of Soviet mechanization. Indeed, the 1950s was the era in which the United States leaped into the age of automation, and industrial processes in forests and enterprises became mechanized and automated.[22] According to Bowles, the mechanization of wood harvesting in the United States had been developing for a few decades while the Soviets expected to achieve advanced levels in just ten years.[23] While biased by the Cold War binaries, Bowles was right to emphasize a typical feature of Soviet industrial policy: the regime always tried to make a leap in a short time by means of intensive technological modernization. As one front-page article stated, "Any capitalist country will require decades to go through the same path that the Soviet forestry industry has gone in terms of technical re-equipment and introducing new techniques and technologies of wood harvesting."[24]

Despite claims about the ability to make rapid progress, the alarmist view insisted that these short official time frames were simply not sufficient to produce a competitive industry with numerous operations. The lack of mechanization was a constant problem facing Soviet wood harvesting, not only due to a lack of funding, but because of rapid technological transformations in wood-harvesting machinery in Western countries. Thus in the late 1940s and 1950s, the chain saw, including the famous "Druzhba," was the main instrument used for felling trees, while the use of modern tractors and giant scissors was spreading in other countries.[25] This and later models of saws were excessively loud, and due to the high level of vibration, forest workers became afflicted with so-called vibration disease, which impacted workers' hearing abilities.[26] Technological transfer and the importation of machinery that, as with forms of Stalinist industrialization, were taken as the main strategy beginning in the 1930s, were important for the industry. Many wood-harvesting and especially pulp and papermaking enterprises were equipped with foreign machinery and mechanisms. For example, the Soviet-Finnish border enterprises were originally equipped and later modernized mainly with Finnish machinery, with Finland serving as the bridge between domestic and Western technologies across the Iron Curtain.[27] In the 1930s, Finland itself experienced rapid industrialization in its forestry industry, and maintained a leading position in the world in terms of wood harvesting as well as pulp and paper manufacturing. If the United States and Germany served as sources of new technologies for heavy industry during the enforced industrialization of the Stalinist era before the Second World War, small Finland fulfilled a similar function for the Soviet forestry industry during the Cold War. Many Soviet forestry specialists strongly believed in the role of Finnish assistance, despite the conditions of the Cold War, referring to the similar geographic and natural conditions of Finnish and Soviet forests. From the late 1960s, it was Japan that took the lead in supplying the USSR with forestry equipment. In July 1968, the USSR and Japan signed an agreement for the development of the forestry industry, according to which Japan was to supply machinery, materials, and goods while the USSR supplied raw timber as the means of payments.[28] Japanese companies supplied, among other things, diesel bulldozers, cranes, trucks, electric cables, and other machinery for Soviet logging enterprises, and received huge amounts of timber in return.[29] This technological dependence revealed the Soviet ability at adapting foreign technologies and lead industrial processes to a large extent with imported machinery.

Despite the imports, some specialists remained critical of the technological level of wood harvesting and processing in the USSR, arguing that the industry was too extensive and diverse to be fully equipped with machinery purchased only from abroad. Many referred to the technological factor and specifically the Soviet Union's weak level of mechanization as reasons for comparative backwardness with the West. For instance, a specialist wrote in 1979 that wood-harvesting enterprises "had been equipped with machinery slowly," and according to his calculations, mechanized tree felling made up a small share of the total harvest in 1977, insisting that "a serious obstacle for technical progress in the [forestry] branch was the absence of modern basic automobiles."[30] The lack of technical infrastructure was therefore a condition of falling behind modern levels, which specialists measured according to the West—the benchmark of standards of forestry development. In addition, due to the lack of qualified forest workers, foresters called for local dwellers to help reforest cut territories, indicative of a typical Soviet practice of using citizens as an additional standing workforce.[31] Soviet forests were the main work sphere for prisoners at the infamous concentration camps (Gulags). Prison labor was still practiced, while significantly reduced, in the three decades after de-Stalinization was launched in the mid-1950s. Wood harvesting remained a difficult job, mainly done by seasonal workers, which included female laborers due to the postwar lack of male laborers. A Finnish engineer, who was intensively engaged with the cooperation with Soviet engineers and once visited the Soviet Union in the 1960s, was surprised to see how Soviet women cut trees in the winter wind, standing in deep snow.[32] In some regions of intensive forest exploitation, over 35 percent of the total workforce was made up of women, who usually worked with harder and less sophisticated operations.[33]

From the 1950s onward, however, there were some attempts to improve the condition of work in forests to increase the productivity of wood harvesting. The government and specialists increasingly emphasized the importance of making the conditions of wood harvesters more comfortable—in particular, by introducing better workers' clothes, shoes, and safety gear. Existing shoes and clothes were not reliable as most forest workers wore artificial leather (*kirza*) boots, which quickly became wet and torn, while in winter they wore felt boots (*valenki*), which easily became wet and shrunken. At that time, workers in some other countries wore heat-retaining rubber shoes and safer warm hard hats.[34] In the USSR, the problem of workers' clothing was not completely solved until at least the late 1970s. This prevented new and young workers from entering the industry, which remained hard seasonal work.[35]

Beginning in the 1950s, some industrial factory specialists discussed the quality of supplied wood to enterprises more frequently, addressing wood harvesting as a critical problem. Industrial enterprises regularly complained about the low quality and shortage of raw materials when explaining the work stoppages typical of state socialist enterprises. According to the experts' assessments, the quality of wood posed a serious problem for industrial enterprises as it was often in a state of decay and thus inappropriate for processing.[36] If archival sources are to be believed, this problem was also recognized at the highest political level. The Council of Ministers, the executive center of the country, growing concerned with the lack of improvement in forestry practices in the Russian Republic, decreed that approaches to forestry had not been effective, especially in the European part of the country, where "intensive cuttings led to the depletion of forest."[37] Despite numerous problems related to technological infrastructures and the workforce of forest work, the Soviet wood-harvesting industry was one of the most voluminous in terms of production numbers.[38] As one specialist calculated, measuring harvests with final consumer products, the USSR harvested enough wood per minute to fully furnish 210 two-room apartments.[39] Yet at the same time, in 1962, the USSR produced five times less paper than the United States, while the gap in making cardboard was even bigger.[40] Between the 1950s and 1980s, the tempo of papermaking in the USSR was one and a half times quicker than that in the rest of the world, yet there remained a large gap in production when compared with the main Western producers, such as the United States.[41]

Harvesting enormous quantities of wood yet manufacturing relatively few consumer products, Soviet industry exported a great deal of timber abroad; in 1964, timber was sold to fifty-two countries.[42] A significant proportion went to neighboring Finland, which had a well-developed wood-processing industry; from 1965 to 1966, Soviet timber exports to Finland doubled. At that time, Finnish companies were allowed to transport timber to Finland directly from logging spots. One Soviet propaganda film proudly claimed that this change "enlarge[d] our trade connection with friendly Finland."[43] It presented increasing exports of raw materials as a positive sign of developing bilateral economic relations. Yet in fact, it signaled large-scale losses of valuable raw materials for Soviet industry.

The lack of wood for supplying enterprises originated not only from wasting harvesting and imports. Timber was also lost in the process of floating and drifting logs down waterways and rivers—the cheapest way of transporting wood in the USSR until the mid-1980s. In the 1930s, up to 3 percent of timber was lost from floating—a number specialists regarded as significant.[44] Roughly the same levels of loss persisted in the late 1970s, despite specialists having long concluded that timber drifting was a wasteful and environmentally dangerous practice.[45]

Overall, the gap between the volumes of wood harvesting and processing led some specialists to express disappointment about the performance of the Soviet forestry industry, describing it as being at a critical stage. While forestry had been wasteful before, the strong demand for wood-based consumer goods and rapid deforestation all intensified the image of a coming crisis. Specialists saw this crisis stemming from technological and infrastructural backwardness, and warned that it would lead to wood shortages. If industrialists blamed wood harvesters for supplying enterprises with wood of poor quality, harvesters in turn pointed to the low capacities of the wood-processing industry due to the technological factor. As some wrote in the last years of the Soviet epoch, the output of Soviet forestry, pulp, papermaking, and wood-processing industries still did not allow for the effective use of the Soviet Union's numerous forest riches. It could not "satisfy the needs of the national economy in most types of paper products. Possessing the biggest forest resources, our country significantly lags behind a few developed countries."[46] For specialists, crisis implied the impossibility of meeting economic demands, and this posed a danger to forests because of the low productivity associated with their long-term, extensive, and inefficient exploitation.

Some stated that the problem of wood harvesting lay in the fact that planning was based on the existing infrastructure rather than on the availability of forests. This is why, they said, the northwest, where historically technical infrastructures had been longer developed, were heavily devastated while the greener covers of Siberia and the Far East largely were terra incognita for harvesters because they remained inaccessible. They also complained that in the postwar decade, the industry had adopted "the principles of the long distant 1930s when sustainable forest use was anathematized as something not corresponding to the program of enforced industrialization of the country."[47] The postwar period revived many approaches and ideas of the 1930s, most of which were seen as positive and fruitful. At the same time, however, many criticized these old practices, claiming that the new age required new methods and instruments. As ministry specialists argued, the rapid industrial development of the country in the prewar period and restructuring of the Soviet economy after the war led to a situation in which the forestry industry was supplied only with round wood, with little invested in sophisticated projects for transforming raw wood into modern goods.[48] The focus on industrialization and the rapid use of forests that had been emphasized from the 1930s on and remained widespread thus attracted criticism from forestry specialists. As one professional report on the development of forestry in the 1960s insisted, there was a "*careless* attitude toward our forests.... Logging companies have neither [production] stimulus nor material motivation to rationally use . . . even available wood stocks, they also do not have any stimulus for introducing scientific achievements in the forestry production."[49] Forestry specialists held some power to complain publicly about the low levels of investment in industrial forestry and criticize the state of things in the industry.

Overall, the period from the mid- to late 1950s became an important era for acknowledging the critical state of wood harvesting, both by wood harvesters and producers at industrial enterprises. This critical juncture in thinking about wood availability and practices of harvesting made many seriously rethink the interplay between the geographic distribution of forests and their industrial exploitation. Wasteful wood harvesting, backward wood processing, and the insufficient manufacture of modern goods together constituted the components of the critical language of the professionals. Importantly, specialists did not use the word *crisis* themselves until the end of the Soviet regime, but nevertheless used the language of crisis and explicitly expressed an expectation of the critical state of Soviet forests. By the 1960s, specialists recognized that the wood-harvesting industry was at a key point of inefficiency, careening toward wood depletion.

In most alarmist responses to the expected wood crisis, backward technology was given a crucial role in explaining the low levels of harvesting and consumption of wood. By the 1980s, however, more specialists connected the danger of rapidly disappearing forests with a human factor, insisting that forests were an important natural *actor* rather than merely a natural resource. As engineer V. Shiryaev wrote, "Today many think about foresters as barbaric people who deplete a national treasure—forests. We indeed have evidence of that: wasted forest sites, decayed wood on the roadsides and river banks, drying lakes. After decades we have to admit a crisis in the forestry complex. Forests are depleted, and machinery equipment is used in a very ineffective way." He found the reasons for this critical state not in technology as such, as specialists had previously argued, but instead in human activities—when workers in forests and specialists did not want to do their work carefully because most capital expenditure went to the upgrading of machinery and production processes as opposed to people and their living standards. This was why, he said, we developed "a barbaric attitude" toward forests, rivers, and land.[50] The human factor became central from the 1980s on, in tune with broader state appeals to the human face of socialism in the course of perestroika. It was also connected to a growing environmentalism in which more voices spoke loudly against pollution and environmental degradation in the country. This exemplified the transformation of concern over the destiny of forests, from the hidden reference to humans as a destroying force through criticism of the overexploitation of European forests to the explicit criticism of human activities, showing the strands of environmentalism found in the industry.

By the end of the 1980s, sustainable forest use was evoked as the counterforce to resource scarcity in wide discussions among Soviet specialists. They particularly criticized the Soviet forms of managing resources when enterprises were to control the availability of their resource base themselves in most parts of the Soviet Union (except at forest-industrial complexes [LPKs]). As calculated by some, of twelve logging spots (*lespromkhozy*) in Karelia, one of the most intensively exploited forested regions in the Russian northwest, eight would no longer be capable of working in the near future because of the depletion of wood stocks in the region—the result of which would "lead to the destruction of enterprises and serious social consequences." By the year 1990, a "condition of serious deficit of wood resources" was described in this region.[51] If in earlier decades specialists warned about the *envisaged* wood crisis because of some wasting practices, in the last decade of state socialism they emphasized *real* failures in wood exploitation. Where at first the imagined crisis emerged from various industrial processes that attracted criticism from specialists as ill-suited to the modern time, it later turned on the recognition of humans as a destructive force.

PERPETUUM MOBILE OF ADMINISTRATIVE CHANGE

The Soviet forestry management system of the central state and its regional echelons were the subject of constant reorganization. Reform and structural change were held out as the administrative solution to the wasteful development of the forestry economy. The ministerial enterprises (i.e., sawmills and factories subordinated to the Ministry of Forestry) were the main consumers of wood in industrial operations. The ministry had, however, changed its names and subordinations many times and suffered from these continual organizational transformations, which emerged from the recognition of crisis in managing forests and wood resources. In 1957, after the administrative reform initiated by Nikita Khrushchev, the ministries were dissolved—yet the forms of territorial administration introduced in their place did not make the system of wood harvesting and processing more effective. *Sovnarkhozy* or new territorial administrations were made fully responsible for fulfilling the plans on production, supply of wood, and developing new technologies. Because of that, the localization of new logging companies—lespromkhozy—often led to the rapid depletion of forests and prompted foresters to move to new zones, constructing expensive infrastructures and production units there. To some extent, this lead the Soviet type of harvesting to follow a destiny similar to that of earlier and contemporary histories of many other places in the world where forests were cleared entirely, ranging from the eastern parts of North America to the Amazon River basin.

From the 1960s on, after the failure of the sovnarkhoz reform that aimed to organize the territorial administration of economic development, the government invested in so-called large enterprises as a new form of governance that was frequently called progressive and efficient. The gigantomania of Stalinist industrialization, when large-scale enterprises were built as a sign of rapid industrial construction, was now supplanted by the gigantomania of *management*, when enlarged administrative bodies were considered to offer the most efficiency compared to small enterprises directly subordinated to the ministry. By the 1970s, the state constituted industrial self-supporting associations (*khozraschetnye ob'edineniya*), including furniture making, sawmill and wood processing, and match-making associations. These were large industrial complexes subordinated to the ministry. For example, seventeen sawmill and wood-processing associations were comprised of more than four hundred enterprises and organizations, altogether employing more than four hundred thousand people. Some specialists looked on these changes positively, insisting that they helped solve the problem of the management of separate branches of the forestry industry.[52]

In practice, this constant reorganization and administrative gigantomania led to numerous changes that complicated the activities of the enterprises. The Lyaskelya papermaking factory, a former Finnish enterprise that had moved to the USSR after the Soviet-Finnish War in 1944, serves as a good example of this dynamic. In 1958, it was united with the cellulose and papermaking plant in Harlu, a settlement located nearby. Between 1944 and 1951, both were subordinated to the industrial association of the papermaking industry and then moved to another until 1953, only to be subordinated to the previous association once again until 1955. It then moved to the cellulose-making association until 1957, and later to the Karelian sovnarkhoz as part of the sovnarkhoz reform. In 1961, the enterprises were reformed into the Lyaskelya cellulose and papermaking plant, which changed its subordination seven times. Multiple reorganizations prevented this factory, previously a prosperous Finnish enterprise, from playing a role of any real importance in Soviet forestry production. Instead, the factory was repeatedly described as outdated, nonmechanized, and quite literally bankrupt. Indeed, by the late 1980s and early 1990s, it was closed and finally abolished along with the socialist regime in Russia.

As with wood processing, various operations of wood harvesting were distributed between many organizations. A small sector belonged to hunting companies and nature reserves, the Ministry of Transport, and other ministries and economic institutions. Quite a significant proportion of forest belonged to kolkhozy and sovkhozy, the Soviet type of collective farms. This division was fairly conventional, though, as quite often there were several similar organizations that "possessed" forests in the same region. Moreover, the Ministry of Forestry counted about two thousand geographically scattered enterprises. As a result, "logging companies were distant from central production spots [by] thousands [of] kilometers, while large structural institutions supervised territories of hundreds and sometimes thousand [of] kilometers."[53] Obviously, extreme distances and rapid cuttings led to wood-processing enterprises being located further and further away from logging spots. This created complicated infrastructure that exacted huge expenses on logistics to deliver appropriate types of wood from logging spots to the industrial wood-processing enterprises. Specialists typically complained about the gap between wood-harvesting and wood-processing operations when two processes of the same industrial chain became separated from each other. They described how in the course of doing their job, harvesters did not really think about how to use the wood, while specialists working at wood-processing enterprises often blamed harvesters for industrial problems, such as low volumes and the poor quality of ready products. Another serious problem from the industrial perspective rested with the fact that some forests were declared protected territories. As industrialists complained, these forests became old and could not be used industrially, so they did not produce economic benefits.[54] Due to this, many recognized that intensive cuttings in the European part of the country led to the "depletion of forests." They also stressed that poor management and an insufficient safety system led to numerous forest fires and the spreading of pests that damaged trees.[55] In turn, logging companies complained about the absence of funding for reforestation. If these complaints are to be believed, by the 1970s in the northwest, the volumes of wood harvesting were disastrously high and led many to think about decreasing rather than increasing the rate of cutting, thereby breaking the logic of Soviet industrial planning that aimed for ceaseless increases in production. Most important, this gave added impulse to alarmist views and expectations about the future of forests, which saw them as a space where industrial demand and nature competed.

Warning about the prospect of wood scarcity, specialists frequently complained about interagency obstacles, the lack of expertise, and the absence of working plans as reasons for numerous problems in the industry, such as the clear-cutting of industrial forests. Some logging companies and enterprises often did not fulfill the plan. Forest species that had been demanded just perished in forests and were thrown away due to deficiencies in planning and transportation problems.[56] Sometimes, forestry harvesters did not complete the clean-cuttings needed for reforestation, and the task of keeping forests clean and healthy was obstructed by the territorial dispersion of trees along with the absence of forest roads to them. In Karelia in the late 1950s, cuttings were most intensive in the southern and western areas due to the availability of railroad transport, yet were no less intensive in other parts of the country.[57] One forestry inspection noted that "because of the lack of clean-cuttings in the forests of Karelian Peninsula . . . we observe the large-scale *dying* of trees. . . . [F]orests are wasted, and forest fire danger increases."[58] Professional observations often anthropomorphized forests with metaphors such as living and dying (and sometimes dead) species. Dead forests in their language usually implied perished forests, but wood was associated with the technology-enabled continuation of the life of the forest. Official documents and specialists' reports ritually blamed local forestry officials for their "careless" attitude toward forests, frequently describing the forests as perishing organisms because of the lack of proper care. Particularly by the end of the Soviet epoch, the lack of care and overexploitation, in their view, had led to the destruction and damaging of economically useful species. This depiction revealed how specialists related the living cycle of nature to industrial purposes along with its economic service for the society and regime. Care and attention in this sense implied the proper economic use of forests—strands that would come together and develop as a discourse of industrially embedded ecology within the industry.

Specialists recognized that the forestry industry as a whole was in crisis, wracked by practices they described as low in productivity, wasteful, and weak. This, they believed, would lead to wood scarcity and the depletion of industrially appropriate forests in the future. Technology played a peculiar role in these explanations, casting the impending wood crisis as a matter of poor technology: technology had performed badly in the northwestern region, where it enabled intensive clear-cuttings; the absence of technologies and infrastructures to exploit the heavily forested eastern lands also contributed to this expectation of crisis. An alarmist expectation of the coming shortage of wood supply recast nature as a finite resource, contradicting the long-held image of green abundance and triggering further thinking about how to keep it from depletion beyond just administrative reforms. The constant reorganization of forestry management demonstrated the government's recognition of the crisis of wood harvesting and other forestry industry branches from the economic and technological perspectives too, as specialists did.

Apart from institutional changes, which obviously had little real effect on the development of the forestry industry, there were other solutions suggested by specialists as significant alternatives to state-led institutional reorganization. Among these was the imperial initiative of expanding wood harvesting into the eastern parts of the country as part of large-scale advancement to Siberia and the Far East, which many saw as a possible solution for numerous critical problems. As specialists believed, extensive advancement would have to be combined with rational and intensive methods, finding the necessary instruments to overcome the wood crisis and forest depletion. This would simultaneously promise more efficient industrial use and a safer future for the Soviet Union's green stocks. And at the intersection of economic interest and finite resource availability, industrial ecology would emerge as a form of care about the future of natural resources.

2

THE INDUSTRY EXPANDS INTO SIBERIA AND THE FAR EAST

THE TABULA RASA OF THE SOVIET FOREST MAP

While in general, forests were framed in the industrial imagination as part of a Soviet commonwealth, Siberian and Far Eastern forests were held to be an even more important and promising treasure as underexploited green covers. As the minister of the Soviet forestry and wood-processing industry between 1968 and 1980, Nikolay Timofeev enthusiastically wrote in 1979, "Siberia and the Far East are the future of our industry."[1] In the language of specialists, new technologically led colonization of the eastern green lands was seen as a possible solution for preventing wood scarcity of the kind that afflicted the depleted northwestern forests, thereby making forestry production more efficient. The hopes placed on the eastern parts of the country derived from the long-term historically developed image of their resource abundance, rooted in ethnographic research conducted in the nineteenth century. Traveling to do research in this region, ethnographers of the Russian Empire produced overviews and guidebooks on Siberian landscapes giving deep details of its natural resources. Describing the Yakut people along with their cultural and natural environment, for instance, one book wrote that "close to settlements [of the Yakut people] taiga is hardly accessible: trees grow widely and quickly, if a tree falls down, it rapidly gets covered with moss and young underbrush. This taiga is groom, there are no berries, no grass, no birds, only the forest makes noise and makes a traveler scared."[2] For Russia, the eastern lands were a frontier, as in the case of the US frontier, calling for investigation.

This image was kept alive after the Second World War and confirmed by specialists, who actively engaged Soviet economic interest in eastern natural resources. If professional publications are to be believed, the Far East forest reserves constituted over 40 percent of the entire industrial forest stock of the country.[3] While the old harvesting region of the northwest seemed to be advancing toward a wood crisis, the new densely forested but difficult-to-access regions of the east of Russia attracted Soviet foresters, promising potential salvation from wood scarcity and prompting greater urgency for the expanded industrialization of these forests. The importance of using these stocks economically was especially articulated in a series of conferences held from the mid-1940s on by the Academy of Sciences and the Council on the Study of Productive Forces, the organization tasked with searching for more effective ways of using strategic natural resources. Such meetings usually issued resolutions that summarized the discussions, and as one stated in 1962, the Amur region in the Far East particularly "has *endless* opportunities for the development of productive forces." This as well as other new territories were conceived as an endless resource for building consumer industry, or in the specific Soviet language, the "material basis of Communism." The meeting criticized geologists and other scientists for "lagging behind" in the exploration of minerals and other natural resources, which as these scientists expected, were meant to serve the country rather than lying "homeless" in nature and making for an economic loss.[4] This industry-oriented view pointed to the uniqueness of the Far East, emphasizing that the region was the second-largest forested territory after Siberia and exceptional in terms of the number of valuable tree species growing there.[5]

In accordance with the Soviet tradition of linking state achievements with the activities of Vladimir Lenin, Soviet publications often invoked him as the figure who originally suggested locating industrial production near logging enterprises to minimize transportation costs. In fact, the late 1940s and accelerating drive for technological modernization in later decades was marked by a radical increase in interest toward new forestlands. In 1960, the journal *Lesnaya promyshlennost'* (Forestry industry) not only described the Soviet Far East as a place of enormous and versatile forest riches but also argued that they were industrially valuable natural resources that should be harnessed to serve (*sluzhit'*) the economy. It thus saw the underexploitation of forest riches as a drawback given that nature had to serve economic interests, which at the time were framed as the interests of "the people." The article boasted that more than a quarter of the whole Soviet fir and spruce stocks were concentrated in the Far East—trees considered to provide "the best raw material for making pulp, viscose, and highest quality paper."[6] Specialists perceived the possibility for advancing to the Far East as urgent and encouraged strong enthusiasm among harvesters in the forestry industry, supported by research that gave a much clearer picture of the quantities of wood stocks that were available in the 1950s and 1960s.

When almost twenty logging spots were opened in Siberia, specialists calculated that their stocks would be sufficient to provide wood to Soviet chemical enterprises for up to a hundred years. They hoped that these lands would become centers for the rapid and formidable development of the forestry chemical industry in particular—a promising industrial branch—and offer a boon to the search for new sources of energy.[7] Their enthusiasm was largely based on the fact that new enterprises were surrounded by forestlands and could be supplied, at least for some decades, from the nearby territories. This vision of resource use without depletion over the course of many years seems to have persisted in both the USSR and postsocialist Russia. It resonated with how some engineers saw the Baikal waters as a source of pure water for a new pulp enterprise without consideration of potential serious damage in the long term.[8] This continuity was also revealed by contemporary Russian politicians. In the 2000s, the president of Russia sometimes mentioned the possibility of exploiting oil for a few years with no serious concern about alternative sources of energy.

Inspired by chemical achievements across the globe, many scientists working in the forestry industry wrote glowingly of the magical potential of "eastern trees" as a material for growing consumer production and advancing socialist modernity. In the 1960s, at the peak of Soviet passion for chemistry as well as the intensive applications of chemicals in agriculture and industrial production in Western countries, one specialist wrote that due to the specific nature of Far Eastern wood, it was possible to produce fir vitamin powder for cattle.⁹ Others stressed that Siberian cedar on Altai was a unique material that could be used industrially within the region to decrease the need for the transportation of wood from other areas.¹⁰ This economic perspective was important for maintaining excitement about the use of eastern wood, not just because of the anticipated lack of wood in other regions, but because of the very qualities of Siberian and Far Eastern wood.

Planning for the growing consumption of mass products, specialists saw Siberia and the Far East as a crucial source of the raw materials needed for a technological leap from the past to the future. The expansive economic move to the east undertaken by the state was met with strong enthusiasm from both industrialists and alarmist forestry specialists. Many saw the eastern forests as the key means of preventing a wood crisis in the northwest of Russia. While both expectations revealed a strong desire for industrial development, they were primarily connected to the aim of cutting less in the northwest because they connected both parts of the country—the more forests harvested in the east, the less the forests of the northwest would need to be exploited. The central professional journal *Lesnaya promyshlennost'* argued that the expansion of wood harvesting should happen through intensive cutting in the eastern parts of the country while simultaneously decreasing cuttings in the northwest.¹¹ The complexity of this approach lay in the conjunction between the industrial and alarmist lines. Specifically, relocating the capacities and cuttings to the east implied the possibility of decreasing cuttings in the northwest in order to prevent the catastrophe of depletion and solve the problem of raw material supply for the industry.

Given the fact that Siberia was a frontier of state-led colonization in the nineteenth century, it offered a useful historical symbol for framing the Soviets' internal colonial mission. The postwar colonization of eastern forests drew on more sophisticated technologies than had previously been available—tractors, new cutting mechanisms, and so on—even though they were, as many specialists put it, less advanced than those in Western countries. The initiative promised to provide wider economic possibilities while employing what many specialists believed would be a more careful attitude toward the forests. Importantly, Siberia and the Far East were to become the sites not simply of a shifted forest frontier but instead a completely new technological space of resource exploitation to avoid the mistakes and problems of overexploitation experienced in the old forestry region. For specialists, it was typical to compare Soviet economic performance with the past and present, especially emphasizing the technological levels of the czarist epoch and contemporary Western countries. Setting the latter as reachable goals was important for the Soviet system in motivating internal technological development, and such benchmarking comparisons were often reproduced by specialists. They particularly conceptualized economic development through the denial of the prerevolutionary era and measuring standards according to Western (US and European) levels of production. Many harvesters and industrial engineers criticized the previous development of "backward" industry, which they typically connected to a predatory czarist past along with the weakness of early Soviet harvesting and industrial forest use. Soviet professional authors—engineers and scientists—presented the pre-1917 era as having been characterized by the cruel stealing of forests and other natural resources by imperialists. At the same time, they criticized czarism as shortsighted because while it overexploited the forests of the European part of Russia, it underexploited its eastern zones. They argued that this approach, which left huge green lands untouched, was absolutely wrong.[12] Soviet publications praised Soviet achievements and the "progressive" vision of the future, proclaiming that unlike the lazy and technologically poor czarist regime, Soviet power would introduce advanced technologies to forests.[13] This was the broad direction of Soviet policy in wood harvesting before the mid- to late 1940s, when, as shown in chapter 1, economic growth and increasing consumerism sharpened concern about an impending wood crisis. By blaming czarism as the brutal destroyer of the northwest and ignorer of the east, specialists conceptualized the exploitation of forests and advancement to "orphan" (*besprizornye*) forests as key developments along the path to a civilized, industrial world. Compared to the previous Russian colonization of the east, which was moved primarily by economic, cultural, and later political motives, this was heralded as a largely scientific and technological advancement: science and technology, with their material tools and principles for the organization of wood harvesting, were seen as critical instruments to help domesticate new forests and put them in the service of the national economy.

Under the Soviet regime, the volumes of wood harvested in the east increased by sixteen times, and most of the east's industrial capacity was reached in the postwar period. The mid-1950s were marked by significant enthusiasm for not only relocating cuttings to Siberia and the Far East to supply old industrial enterprises of the northwest but also building a network of new forestry industry enterprises there. These would now be based on new rules, more rational calculations, and better infrastructures, which ostensibly meant that the forestry industry would be inherently different from how it had operated before. These expectations were centered on improving the quality of cuttings and industrial processes through the power of technology. Providing huge resources and implementing sophisticated technologies to cut, transport, store, industrially prepare, and transform wood, Siberia and the Far East in this sense were to become the emblems of a new type of forestry practices.

Specialists often referred to rationality and saving costs as instruments to further the exploitation of forest resources as well as the development of forestry, wood processing, and paper industries. These, as analysts expected, would enable harvesters and producers to increase the productivity of forests and improve their economic performance, minimizing losses arising through waste.[14] Many wanted to use new regions as a kind of tabula rasa to build new advanced enterprises, employing new methods and technologies of wood harvesting and processing. As one engineer wrote, "It was much *easier* to do [harvesting] in *virgin* taiga of Siberia and the Far East" than in the *damaged* old region.[15] New lands were to provide raw materials to fulfill this aim and conduct technological experiments, offering "a real basis for the complete satisfying of needs of the national economy in wood."[16] Foresters, including alarmist ones, hoped that the expansion would amount to more than simply a devastating geographic move. Instead, they supported the idea of using alternative resources and waste in order to prevent the wastage of forests typical of the previous practices of harvesting in the northwest. Many scientists and engineers warned that it was important not to follow the path-dependent line and save Siberian forests from the fate of the northwest European part of the country, as they were to play increasingly significant roles in future economic development and thus should not be devastated.

While advancing into new forestlands was a matter of professional hope, it was also part of a crucial political campaign launched by Khrushchev's administration in 1956, tied to key economic aims. In order to exploit huge resources, many forest specialists and officials believed that new technological infrastructures and industrial enterprises would need to be constructed in Siberia and the Far East.[17] The forestry advancement to the eastern forests was part of a bigger campaign for colonizing Siberia and the Far East, and echoed the virgin land campaign unfolding in the Soviet Union at that time. Most of eastern green lands were located far from industrial facilities, and because they were largely inaccessible, were metaphorically conceived as untouched resources. Advocates of the advancement strategy argued that in many parts of new forests, wood harvesters and industrialists were to first tap economically untouched riches, civilizing them and making them economically productive. In the spirit of the time, and echoing Khrushchev's widely publicized unspoiled lands campaign in Kazakhstan, professional forestry publications described Siberian and Far Eastern forests as "forest virgin land." As with depictions of the southern lands for growing corn, they portrayed Siberia as a land where wood harvesters were to journey, like adventure seekers, equipped with novel wood-harvesting technologies. In this picture, harvesters were groundbreakers who did not fear difficulties and were ready for hard work.

There was an obstacle, however. While insisting on the eastern abundance, specialists emphasized one crucial problem with the new forests: they were attractive but hardly accessible and required the deployment of sophisticated technologies. On the one hand, enormous forest resources attracted both wood harvesters and industrial producers, promising great economic opportunities and decreasing the burden on the European region. On the other hand, the real possibility of exploiting them remained quite limited: it required a set of advanced technologies and developed infrastructures that were not often available in the planned economy.

HARD ROADS TO ABUNDANT WOOD

For foresters, it was extremely expensive to move to distantly located and densely forested areas that were marked by an absence of roads and other basic infrastructure. This was made even more arduous by the low level of technological equipment of the Soviet forestry industry in general. While the exploitation of forests in the east was seen as crucial, harvesters were only able to begin to advance to the dense forests after the expansion of oil and gas exploration as well as the advancement of railroad construction.[18] The railroads were particularly decisive in terms of the transportation of wood. While the practice of floating felled trees along rivers and lakes continued until the end of the Soviet Union, it gradually decreased as its role in devastating nature was increasingly recognized. Railroads were seen as a more technologically progressive, environmentally sensitive, and cheaper means for transporting wood. By the 1970s, in many Western countries, railroad and automobile transportation was seen as the most modern form, signifying how technology rather than nature (rivers and horses) could help transport cut trees. By this decade, some specialists considered floating, especially the drifting of wood, as negative for the environment, and emphasized the need to develop roads instead of wasting rivers and lakes, which led to the loss of fish in particular. Other industrialists, however, insisted on the continued development of floating by rivers as a cheap way of transporting wood. Thus by the 1970s, the practice remained a point of intense contest. As one engineer said, there were about 180 rivers in the Irkutsk region alone, which could definitely be used as infrastructural support for forest exploitation in Siberia.[19] However, by the mid- to late 1970s, floating, especially wood drifting, was decreasing, and stopped entirely in some water basins that were recognized as being in danger because of wood bark, sunken logs, and other waste left in the water. This decrease, though, mainly happened in the old industrial regions of the northwest, such as the basin of Lake Ladoga, the largest lake in Europe.

The building of the famous Baikal-Amur Mainline (BAM), which began in the late 1930s and continued in the 1940s on, made it possible to organize not only transport infrastructure but also equipped advancement to eastern forests. BAM in this sense was more than just a grand Soviet project of railroad construction; it was the trigger for hopes of the development of forestry and a solution to the looming wood crisis. In some regions without water basins that could be used for floating, railroads offered the only means for wood transportation. The Upper Pechora railroad, for example, was crucial for transporting timber because the rivers there flowed upward in a northerly direction and could not be used for transporting timber to enterprises located further south.[20] Specialists praised the BAM project for enabling "the favorable conditions for huge industrial exploitation of forests" in the new region.[21] In 1969, the highest political echelons of the Soviet Union, the Central Committee and the Council of Ministers, decreed the exploitation of oil and gas in Siberia, politicizing these two natural materials as principle economic resources. For specialists, producers, and industrial managers, the reallocation of wood harvesting into the eastern regions was intended to supplement fossil fuel excavation. For wood harvesters—in the first instance, at the ministerial level—forests were riches and inherent participants in the forging of a large economic complex in these lands. In this sense, there was a direct link between oil, gas, and wood as industrially demanded modern materials whose *complex* exploitation made economic advancement possible. The government attracted specialized forestry institutions, such as the Institute of Forest and Wood (named after V. Sukachev of the Siberian Branch of the Academy of Sciences), Committee on Forestry and Wood Processing Industry (Giprolestrans), Institute to Project Papermaking Enterprises (Giprobum), and others. Together these institutes fostered a network of activities directed at turning a new page in the history of Soviet forestry.

Inspired by wood availability in Siberia and the Far East, many specialists and planners aimed to develop a network of the modern forest roads needed for transporting wood in the new regions. Historically, poor infrastructure for wood harvesting was one of the main problems for Soviet industry in general. This was especially visible from the mid-1950s on, when automobile road-making technologies had been rapidly expanded in other forested countries while in the USSR it achieved only halting rollout. Soviet harvesters cut large amounts of wood, but the difficulty persisted in transporting wood from logging spots to production units. Enterprises often blamed harvesters for delays and deficiencies in supplies of wood, alarmed by the absence of the raw materials needed for achieving the targets of the plan.[22] Producers also complained about receiving raw materials of poor quality; wood was frequently in a state of decay and wet, or consisted of different sorts of trees, negatively influencing the production process and particularly the quality of cellulose.[23] Enterprises also often complained about the poor quality of wood shipments that derived from a lack of technologies related to the antisepticizing, transportation, and storage of wood. Weak infrastructures in forests and at enterprises prompted some specialists to think about the low capacities of wood harvesting and huge technological burden on nature. These issues may have been exaggerated in sources to justify the failures of five-year plans, but the quality of products indeed depended on the quality of the wood supplied from logging spots to enterprises.

In the 1950s, forestry industry management decided to invest more in the building of automobile roads in forests, but were met with numerous obstacles such as the shortage of materials like stone. These investments proved insufficient, while the costs that arose to advance the project of the colonization of new forests turned out to be much higher than anticipated. As specialist V. P. Tatarinov wrote, road constructors insisted that "roads must be made properly, they should be maintained, and this is why they were more expensive than expected."[24] By the end of the 1960s, most forest roads in the Soviet Union, including those in the new regions, were nonsurfaced; overall, they carried up to 40 percent of the total volume of wood that was transported in the late USSR. Car trucks, which became increasingly widespread in other countries as a preferred method for transporting timber, were insufficient in number, expensive, broke down easily after a period of just three to four years, and due to the poor quality of roads, extremely slow. The building of forestry roads led by the All-Union Project Institute of Forestry (Soyuzpromleskhoz), as some specialists complained, lagged behind the progress of other "forestry industry countries." Hence in 1965, there were just slightly over two hundred kilometers of roads built in the Russian Republic, while in Finland, roads of more modern design and six times this length were constructed during the same year. As specialists in the mechanization of forestry argued, the reasons for the USSR's slow progress lay in the shortage and sometimes "full absence of specialists in road construction in forestry" (which largely remained seasonal work), issues surrounding technical equipment and documentation, and inappropriate construction technologies.[25] In better states, road networks were established in regional centers, but in the Soviet Union they were located in the old regions, far away from wood-harvesting spots; only 20 to 30 percent of whole wood was transported by these roads, and beyond these, "wood trucks ride along ruined seasonal uncovered roads that attach to regional roads."[26] About 60 percent of wood was transported along uncovered and wood-strip logging roads, which were mostly impassable in the spring time and whose viability directly depended on the weather.[27]

By the late 1970s, specialists increasingly highlighted the transportation problem in the forestry industry, not only in the old regions, but in the eastern lands. In a 1979 article titled "Roads Are the Problem Number One," an employee of the Ministry of the Forestry Industry of the USSR, M. I. Brik, stressed that "modern tempos, quality of construction and exploitation of roads do not correspond to the conditions of [modern] wood harvesting."[28] Due to poor road infrastructure, new forests were being explored far too slowly from the perspective of industrialists. The drive for economization led state authorities to halt the development of gravel roads in favor of snow and ice roads—materials offered by nature "for free." This led to the development of seasonal wood harvesting, which in turn resulted in resource shortfalls due to insufficient cuttings (*nedoruby*) and the inaccessibility of forest resources.[29] At that time, numerous substantial transformations in the forestry industries of Western countries led to the development of better technologies of production, automation, methods of cutting, transportation, and wood processing. If Soviet sources are to be believed, by 1981 about 80 percent of forest roads in the USSR were built without the kind of hard cover that had long been widespread in Western countries.[30] This showed the technological gap between East and West that emerged as a perennial problem during the whole Soviet period.

The situation in Siberia was especially complicated owing to the climate factor; indeed, radical fluctuations of temperature, permafrost, and high levels of humidity were significant obstacles for road making.[31] Many specialists argued that because of the geographic specificities of new regions, better-quality hard-covered roads were needed in place of the nonsurfaced roads still widely used in the rest of the country.[32] Another big problem was the lack of interinstitutional communication, which in general was a chronic problem in the Soviet Union. Some specialists wrote that forestry operations in the east were conducted by numerous institutions—an administrative behemoth demonstrative of the large scale of colonization. Yet each used different methods, not thinking whether their own volumes of harvesting wood might be compatible with others. Some forestry roads had not been properly maintained, and as a result, quickly became ruined. Wood was in demand, and a few industries and numerous institutions shared the right for exploiting forests, creating "many unsolved problems that hindered the effective use of forests and reforestation."[33]

In Siberia and the Far East, forests were particularly far from newly constructed railroads, similarly suffering from the insufficient construction of automobile roads. Forests near railroads intended for transporting other goods were often quickly clear-cut, while other dense stocks remained far away. In this sense, the situation in new lands met problems that the Soviet forestry industry had already tackled in the northwest and brought distinct challenges to the hopes of starting a new chapter in forestry. Heads of construction organizations often referred to the lack of road construction techniques and qualified workers, breaking initial promises to create a more ecologically sensitive forestry industry in the new lands.[34] Sometimes harvesters did not transport wood from logging spots to enterprises due to the fact that all-year forest roads were not appropriate for use and simply left the timber in the forests.[35] The distance between the forestry industry enterprises and economic centers arose as the main problem for Soviet wood harvesting. Many explained the poor supply of wood at enterprises in terms of their location as well as what they called the irrational use of wood.[36] It was a rather typical Soviet problem that echoed experiences of much earlier decades; one might recall the Magnitogorsk iron- and steelworks, where nearby raw materials were quickly exhausting.[37] In this sense, the forestry and wood-harvesting industry followed the destiny of earlier Soviet projects of industrializing nature.

In the course of the industrialization of eastern lands, building the rational infrastructures for resource use emerged as a key priority of experts who supported the state's drive for developing big industrial projects. Referring to the problem of roadbuilding and wood transportation that arose from the beginning of the industrial colonization of the new lands in the late 1940s, specialists justified the importance of so-called territorial industrial complexes in Siberia, and insisted that both intensive wood harvesting and transportation—especially the problem of the "irrational use of railroads"—made the establishment of industrial enterprises in the far-distant lands essential to preventing the "wood crisis" in the east of the country.[38] As many believed, new complexes would bring all industrial operations close to each other and ease the problems associated with the transportation of wood in the colonized regions.

SOCIALIST ENTERPRISES OF THE FUTURE

Discussing the role of science in the mechanization of wood harvesting in Siberia, B. Tikhomirov, the head of the Siberian branch of the Research Institute of Forestry, envisaged that "exploring [the] forests of Siberia and the Far East [will be] a big problem of the people's economy."[39] By this he implied that enormous and far-reaching Soviet plans required huge financial and material investments. Indeed, during the whole period of colonization, the state's plans were incredible in their scale. For example, the state decrees issued for the five-year plan between 1966 and 1970 set several monumental economic tasks: to increase oil and gas extracting, establish a few forest-industrial complexes (LPKs), build a railroad between the Siberian cities of Tyumen' and Surgut for the distance of more than six hundred kilometers, complete the oil pipeline Ust'-Balyk-Omsk of almost one thousand kilometers, and finish the construction of several other railroads.[40] Many of these roads went through dense and, as Soviet sources put it, high-quality fir forests. As some wrote, for instance, "From the party's and industrialists' perspective . . . the possibilities of exploration of the richest natural resources of this economic-geographic region allow exploiting [Siberia]."[41] They primarily connected it with the full waters of Angara, a powerful river in eastern Siberia—the only river going from Lake Baikal to provide its water powers for electricity and industrial production.

The plans for electrification in Siberia through harnessing water energy had been discussed since the 1920s. In 1924, for example, a prominent engineer, V. M. Malyshev, suggested that the Soviets build four hydropower stations on the Angara River in Siberia. By the late 1950s, six hydropower stations were projected and partly built in the region, based on which the Soviet leadership developed the so-called Angara-Yenisey territorial-industrial complex that linked together hydropower stations, LPKs, aluminum factories, and excavation sites, among others.[42] This plan was to provide large-scale industrialization for the eastern lands.

At the time, however, most industrial capacities were still located in the northwestern part of the country. By 1967, there were 245 cellulose and papermaking enterprises in the country, most of which were constructed in the northwest during Stalinist industrialization.[43] Among them were the Kondopoga and Segezha cellulose and papermaking plants built between the 1920s and 1930s. As a result of the Second World War, the USSR annexed several additional enterprises with the territories from the Baltic states, Japan, and Finland. Despite the increased capacity of cellulose making after the war, it remained insufficient to satisfy the growing demand for consumer and strategic industries. Between the 1960s and 1980s, several dozen enterprises were constructed in the eastern regions, expanding industrial processes across huge territories through, for instance, the Bratsk and Ust'-Ilimsk industrial complexes, Selenginsk, Amursk, Baikal'sk, and other pulp plants. In the spirit of time, construction in the east was organized in the form of a shock-work campaign. Apart from the famous "all-union construction site" of BAM, there were several more sites to which the government called young people from other areas, hoping to make up for the shortage of labor in the region, with its cold and hostile climate. In general, though, while Siberian forests and waterways saw the juxtaposition of nature and industry, and natural treasure was to be used "by all the means," the "wrong use" of natural riches provoked criticism.[44] There was the infamous Baikal'sk pulp and paper plant as well as the Selenginsk pulp and cardboard facilities constructed near

Lake Baikal, which produced serious pollution and prompted a strong negative response from the public, along with some scientists and even engineers.

Moreover, Soviet industrial logging companies (lespromkhozy) were developed there as a network to provide new enterprises with wood, exemplifying the rapid growth of wood harvesting. In the Far East alone, the volumes of timber increased by 50 percent between 1940 and 1970.[45] Siberia and the Far East were expected to become export sources for foreign countries, and indeed, countries like Britain increased their purchases of Soviet wood.[46] Many specialists believed that through intensive industrial construction in Siberia and the Far East, the harvesting practices in the European part of the country would become less intensive. In the 1950s, a gradual decrease was expected, which in fact took place. For example, in 1932 the European region provided 28 percent of all harvested wood; in 1955, output there decreased to 22 percent, and by 1960 it was set to decrease to 18 percent. By contrast, while eastern Siberia supplied just 3.5 percent of wood in 1932, by 1955, it was producing 11.5 percent of all the harvested wood, increasing to 14 percent by 1960.[47] Problems with accessibility, transportation, and the use of eastern wood, however, failed to entice some producers, which continued exploiting traditional northwestern parts of Russia, particularly in Karelia, resisting the call to relocate their main capacities to the east.

In the last decades of the Soviet regime, industrialist gigantomania and the enforced rapidity typical of Stalinist enforced industrialization was transformed into a passion related to complexity and economization. Thus the urgent pace of building enterprises characteristic of Stalinism came to be combined with an emphasis on the complex use of natural and other material resources to save production costs. These were the industrial imperatives of late socialism that were brought to bear as a solution for transporting raw wood across large distances—from logging spots to industrial enterprises and pulp from factories to papermaking producers. In the mid-1940s, Soviet economic geographer Nikolay Kolosovskiy published a book titled *Industrial-Territorial Conjecture (Complex) in Soviet Economic Geography (TPK)*, in which he developed the concept of the "complex" to define a territorial area consisting of enterprises that combined whole industrial chains in one geographic location. The complex was designed to manufacture products most efficiently and save on transportation costs, the costs of energy, and other expenses—all through the well-calculated and logistically expedient location of connected industrial operations. Kolosovskiy's proposal found its first practical implementation in 1949 with the construction of the Bratsk hydropower station and railroad, set in the midst of dense forests and the Biryusa, Angara, Ilim, and Lena Rivers. Bringing together several production units, it was believed, could help in the intensive exploration of the rich resources of new lands, saving on the costs of the expensive operations needed for the successful advancement to the east. The problem of intensity was articulated by many at the time who believed new forms of production in these industrial complexes could help Soviet industry gain higher levels of productivity.

Since the 1960s, the territorial-industrial complexes (TPK) became the model of Soviet planning, firmly established as the means by which to approach territorial expansion by the 1970s. Part of the Soviet Union's industrial capacity was to be relocated to the eastern regions, and by the 1970s, "the general line of the [Communist] Party in the location of industrial forces was to enhance the development of eastern regions of the country."[48] Many believed that it represented a project to implement rationality, which put as its cornerstone the expression of a progressive attitude toward nature, as this book later examines. Advancement into the new lands and the desire to turn a new page in industrial development—especially in forestry—encouraged many to call for new forms of industrial activity that followed a progressive form of development. Here, *progressive* denoted an approach to modernity that prioritized the transformation of industrial practices through advanced technologies. It also entailed the rational calculation of costs and incomes in industrial building and production. Scientists gave approval to the advancement to unexplored regions and discussed the complex exploitation of Siberia's natural resources at a series of conferences held between the 1940s and 1980s. In particular, three all-union conferences were convened by the Council on Exploration of Productive Forces of the Academy of Sciences on the development of productive forces. The first was held in the Irkutsk region, the second in eastern Siberia in 1958, and the third in 1969 for the whole of Siberia.[49] Experts thus supported the development of the extractive economy, but emphasized the need to save traditionally overexploited forests from depletion.

In the forestry industry, the TPK form was modified to produce the LPK, which represented a hybrid realization of the idea of new industrial production. These were often called "enterprises of the future" (*predpriyatiya budushchego*). The complexes took the form of large forest industry enterprises consisting of several factories that were responsible for not only producing goods from wood but also planting and harvesting wood. A forestry employee poetically wrote that they were "based on organizational-technological unification," which represented the "economic, natural, and socially effective eclipse of forestry, wood-processing, [and] wood-harvesting enterprises and production."[50] Specialists believed that the creation of these complexes was an effective means of economizing, concentrating all operations, from harvesting to papermaking, in one place. Heralding the LPKs as a means of bringing the future to the present, specialists hoped that the new enterprises would process wood masses more effectively by decreasing the logistical costs between the enterprises and logging spots, thereby solving one of the most critical problems of the Soviet forestry industry.[51] They would organize the transportation of wood in a better way and minimize wood waste from logging spots to the places of processing, effectively unifying industrial forestry and forestry production.[52]

To some extent, the LPKs resembled the huge enterprises already functioning in other industries built before the war, such as the Magnitogorsk and Chelyabinsk metallurgical combines. What was new, however, was the production cycle itself: the LPKs were to receive supplies from the same territorial complex and use internally derived waste from harvesting, sawmill, and wood-processing operations in addition to wood. Prioritizing for economic efficiency (in contrast to prewar production) led to the notion of no-waste production: due to the industrial manufacturing of all products in one place, enterprises were to use all of their own resources industrially. New complexes responded to the global post-Fordist economy, which aimed at the kind of quality, lean manufacturing realized at world industrial giants such as the Toyota Corporation in Japan, whose output accelerated from the 1950s and 1960s on. The LPKs' principles implied implementing the ethic of cost saving and attempting to respond to changing consumer demands. Forestry complexes aimed to become enterprises of this mold, developing no-waste production using their own sawmill and industrial waste to manufacture low-quality paper and other products required by consumers. Yet the image of an inexhaustible base of raw materials remained strong among others who believed the LPKs should still depend primarily on the availability of nearby rich raw material bases rather than on methods of sustainable production and resource use. Thus as one main journal in the forestry industry wrote, there was plenty of wood stocks near the Bratsk combine in the Irkutsk region in eastern Siberia—ostensibly enough to provide stable industrial processing for eighty years.[53] Nevertheless, complexity and stable resource supply formed part of the dream of many Soviet planners to create a new economically efficient system of production.

The construction of the Ust'-Ilimsk LPK in the Irkutsk region in the late 1970s and 1980s was another project designed as a socialist shock-work building site. In 1972, the USSR signed an agreement with East Germany, Bulgaria, Hungary, Poland, and Romania to build a new type of wood-processing enterprise. The Soviet government declared it an international construction site, involving socialist partners in the funding of the project as well as the supply of construction materials and equipment, and a labor force constituted by "shock brigades" (*udarnye brigady*). As compensation for their contributions to the construction, the socialist partners received cellulose from the enterprise; by 1985, over 70 percent of all the cellulose produced there went to the eastern economic bloc. Just a few years after the production was launched, though, specialists expressed alarm about the lack of resources nearby owing to depleting wood stocks close to the production site.[54]

The construction of the LPKs also entailed the construction of social infrastructure, which grew into new cities and city-like settlements (*poselki gorodskogo tipa*), along with forestry workers settlements (*lesnye poselki*). The city of Ust'-Ilimsk, for example, devised by the Leningrad Research Institute of Urban Development (LenNIIPgradostroitel'stva), was seen as a modern urban space equipped with contemporary electric stoves, house elevators, and other material amenities. Originally the city was planned for forty-five thousand people, who would live in comfortable multistory houses—residential types that were not typical of many Soviet settlements. Specialists described them as beautifully decorated houses that boasted a sewage system, water supply, radio, telephone, television, and central heating. Services like kindergartens and schools, cafés, libraries, and other social and cultural facilities were built in each city region too. In December 1973, the construction site was declared part of the all-union campaign of the Komsomol Youth organization to attract young people, in accordance with the spirit of Soviet leader Leonid Brezhnev's time.[55] While much of the equipment Ust'-Ilimsk received was from the socialist partners, significant amounts were sourced from France and other capitalist countries, and it thus effectively represented a joint project of socialism and capitalism.

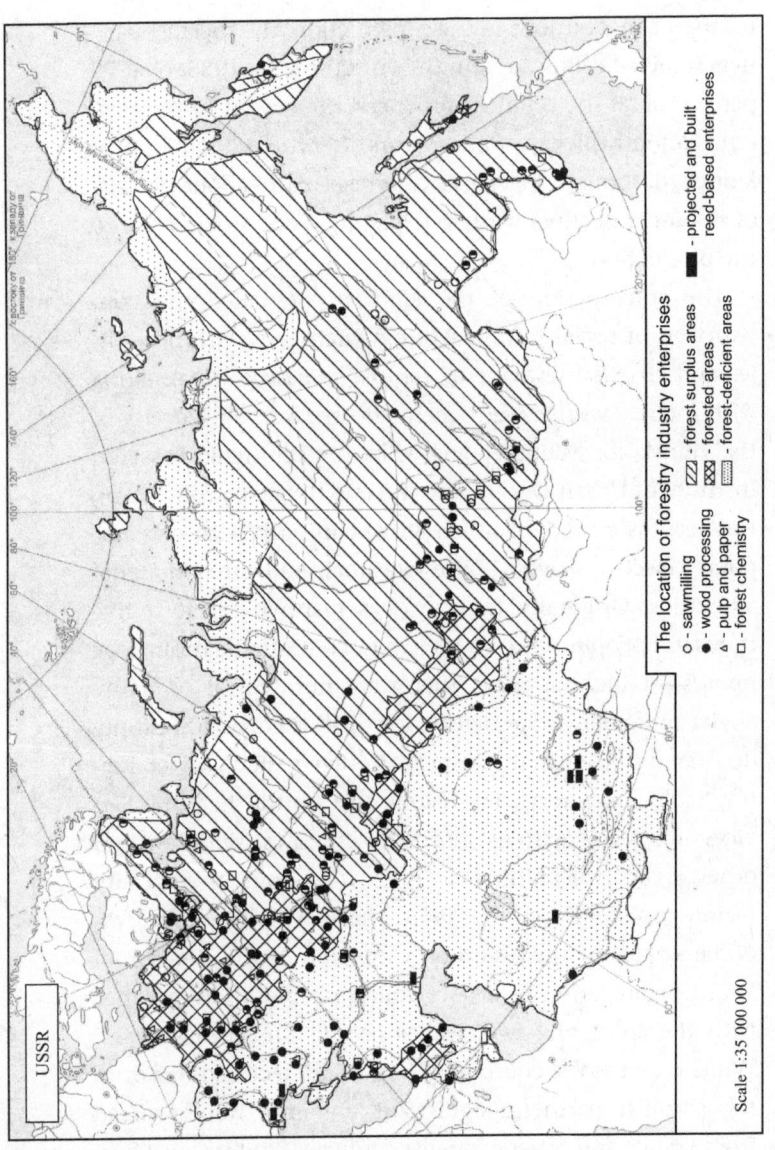

Figure 2.1 Regional distribution of forests and forestry industrialization in the Soviet Union, early 1980s. Data used from L. Kolosova, *Geograficheskiy atlas* (Moscow: Glavnoe upravlenie geodezii i kartografii pri SM SSSR, 1982).

Ideologically, it served as important proof of the Soviet power to build a modern living space in a cold and distant region as part of the LPK. The settlements were to be constructed according to rational planning, with the city space split into functional zones: the residential zone; city center, park, and recreation space; and leisure and greenbelt areas around the city. Specialists described the city as a progressive space with sport facilities and cultural infrastructures corresponding to modern living standards. Ust'-Ilimsk was to be a city near cold water and deep taiga, thereby forging a union between industry, people, and nature that echoed the idea of the "garden city" and "forest city" popular in the early and postwar USSR, as in many other countries such as France and Germany. Overall, half of Siberia's 164 cities were established after 1945, and 40 percent of these were resource based, servicing nearby industries.[56] As in other regions of the country, the forestry enterprises of the east took up paternalistic functions, taking care of the social and cultural life of workers, their families, and servicepeople.

GRAND DISAPPOINTMENT IN THE EAST

Despite large-scale industrial construction in the new regions, many specialists who witnessed the advancement to the east shortly after it began complained about how it was proceeding. Worries about the future of eastern forests were evident already by the late 1950s and 1960s, roughly a decade after colonization was launched. The professional dreams of making a new and efficient form of production of the future, based on the most progressive and rationally employed technologies, clashed with the brute realities of late Soviet colonization of eastern riches. In 1958, specialists at a conference on the exploration of productive forces of Siberia reported that the exploitation of forests was being conducted without any planning for future wood availability and consumption.[57] Some stressed the lack of rational planning, and argued that instead of focusing on industrial building or the construction of wood processing, the state concentrated on wood-harvesting capacities only. As specialist M. Kanevskiy said in the mid-1960s, some 40 percent of Far Eastern harvested wood was mechanically yet not chemically processed, which meant that most wood would not be used for making cellulose but rather was bound for the sawmill. He also complained about the wood waste that was lost while wood and sawmill materials were transported from the Far East to Kazakhstan and Ukraine, entailing the persistence of costly transportation. Kanevskiy insisted that "irrational supplies of timber from the regions of the Far East to the west [of the country] lead to financial loss or the loss of millions of rubles."[58] This criticism of technological slowdown and primitivism in forestry operations served as justification for complexity as an important principle of organizing a new type of wood harvesting and processing in the eastern part of the country. Hopes for realizing the future in the present proved more difficult than specialists initially predicted. Disappointment and skepticism developed in the midst of official hope, with many recognizing that the relocation of forestry capacities to the east required far more energy and resources than at first planned.

While the taiga was the object of conquest, it did not give over its treasures as easily as many initially believed. Conquering new green lands and starting a new chapter in Soviet industrial development were colossal tasks. The promises of rationality, which had served as the ideological base of the eastern project, appeared insufficient because technologically sophisticated colonization required enormous investments. Ironically, opening up Siberia's green landscapes skyrocketed Soviet dependence on natural resources, and the Soviet economy managed to just partly fulfill the initial purpose of the eastern campaign—that is, to decrease cuttings in the northwest while increasing them in Siberia and the Far East. Between 1970 and 1988, the share of Asian regions in the overall volume of cuttings increased from 35 to 41 percent and in the European part it decreased from 65 to 59 percent accordingly.[59] Yet by the end of the Soviet regime, more than 70 percent of paper products were still produced in the European part and Siberia, while the Far East remained poorly investigated. From the perspective of producers, the turn to the east did not in fact represent a radical shift, and many complained that exploring new forests was slow and did not decrease the burden on the old forest industrial regions. Therefore some specialists wrote that "despite a few solutions on relocating the main part of wood harvesting to wood-excessive eastern regions, it is being fulfilled too slowly." According to their calculations, the reason for the lethargy rested with the high cost of constructing new forest industrial enterprises: it was three times more expensive than the reconstruction of already-existing enterprises.[60] The ministry often complained about the price of harvesting in these remote forested territories. As one person calculated, if the industry wanted to harvest wood stocks in uninhabited regions, it would require an incredible sum of almost 8 billion rubles and up to 1.5 million workers would be needed to construct all the railroads required. Others claimed that the initial calculations had been done "by eye" and "someone's will," and were not based on rational research. It was for this reason that the sheer quantity of resources necessary for colonization to the east had been massively underestimated.[61]

In addition, industrialists complained about the slow progress of industrial and social construction, insisting on excessive cuttings and wood compared to real wood-processing capacities. From their perspective, wood harvesting was not productive because the volumes of cut wood were lower than planned while most operations were fulfilled manually; hence in the mid-1960s, more than 60 percent of the workers worked on manual operations.[62] Professional self-criticism brought forth recognition that they were lagging behind the plan and dashing the great hopes placed in the extensive relocation of industrial capacities to provide a rapid solution to the wood crisis. In the mid-1960s, specialists complained that the wasting practices of wood harvesting in new lands were as bad as they were in the old forests. In particular, unsystematic cutting and slash-and-burn methods of cleaning logging sites in the Far East damaged the possibilities of regrowth, transforming forests into deserts.

Gradually, criticism of the industrial advancement to the eastern green lands sharpened, concluding that the colonization experience had repeated the mistakes of previous decades of forest exploitation in the European part of the country. The expansion into new forestlands had several negative outcomes typical of the intensively exploited northwest territories—which many specialists had explicitly tried to avoid. Some said that the main drawback of cutting was that they did not admit the specificities of the region and its forest species. In particular, the rules of cutting (*pravila rubki*) allowed clear-cutting in any industrial forest zone.[63]

Some critics also focused on the social and urban construction in the new regions.[64] They cited poor social and cultural infrastructure along with the lack of comfortable conditions for dwellers at even an elementary level. By the mid-1980s, there were 202 forest settlements (again, lesnye poselki) in the Far East—only 34 percent of which had a supply of water, only 29 percent had wastewater systems, and only 32 percent had central heating.[65] Explaining these problems, engineer V. Gorbachev wrote that the location of lesnye poselki near logging spots and enterprises had not been well planned, was largely ad hoc, and lacked appropriate social infrastructure. He specifically complained that the planners did not consider the principles of settling together with the perspectives of long-term logging activities.[66] As a result, the zone of deforestation and social stagnation extended in tandem: after forests were cut at a particular logging spot, poselki frequently became degraded and left abandoned, empty shells of socialist technological modernity in its colonized zones. Rather than a beacon of progress as they were initially conceived, they in fact became sites of social tragedy on the scene of nature, where social infrastructure was subordinated to the industrial use of natural resources. Social abandonment and degradation was a rapid process; between 1958 and 1959, for instance, about seventy thousand square meters of dwelling space were written off in one region in the Far East alone due to the relocation of harvesting spots. Gorbachev especially complained about the situation in Kalganak, a poselok in the Tomsk oblast' that was constructed in 1957. There, in just six years, all forest resources were depleted, rendering the poselki a senseless settlement. Another example was poselok Balakhtash in the Krasnoyarsk region in western Siberia. Founded in 1952, Balakhtash was quickly surrounded by clear-cut zones because of imprudent planning. There were some additional problems: making well-furnished and comfortable settlements proved expensive; while many saw the living standards of the 1950s and 1960s as much higher compared to the 1930s, people lived in poor material conditions, suffering in favor of intensified industrialization. Another issue was food supply to these remote settlements often established in permafrost zones. The social history of wood colonization fueled continued criticism of Soviet forestry practices.

The Ust'-Ilimsk forestry complex, one of the key 'enterprises of the future', is an example that revealed some of the difficulties that came with the colonization of new forestlands. Despite the Ust'-Ilimsk project resulting in a modern city and nearby enterprise, industrial production there quickly led to the devastation of wood resources in the vicinity, attracting harsh criticism from some specialists. By the 1980s, it became clear that the resource base for the construction in Ust'-Ilimsk was insufficient, and "calculations on raw materials were fulfilled without any unified methodological approach and contradictory to the requirements for sustainable [*neistoshchitel'noe*] use of forests."[67] Even so, the project was continued and enlarged, but it evoked more environmental concern than before.

Ust'-Ilimsk was not the only symbol of disappointment in the failure of economic, environmental, and social sustainability. "Irrational exploitation of the forest stock of the Bratsk LPK," another huge LPK, "led to enormous losses of wood and to the early devastation of raw materials."[68] In the first instance, it meant losses of any type of raw material, including waste, and entailed slow and problematic construction. Numerous ministerial reports demonstrate that problems typical of the Soviet past were once again permeating the new enterprises, including weak labor organization and equipment accident risks.[69] Some sources, such as the 1978 Forest Code (Lesnoi kodeks) devoted to forest legislation, mentioned the sustainability of natural resources as an important principle of economic functioning. From their perspective, it involved not reforestation as such but rather the search for new, rich wood stocks and alternative resources to bolster the sustainability of industrial raw materials. Late Soviet forestry legislation followed the tone of previous decades, when late imperial foresters and early Soviet specialists were concerned about reorganizing forest use to make it more compatible with economic imperatives—emblematized in particular by the 1918 Basic Law on Forests (Osnovnoi zakon o lesakh) and 1923 Forest Code, which stipulated forest protection and reforestation as crucial issues.[70] As the 1978 Forest Code argued, the sustainable and rational use of natural resources was important for satisfying the needs of the planned economy. This is why some specialists explained that it was not right to continue the construction and enlargement of the LPKs, which literally implied the devastation of wood nearby. The project in Ust'-Ilimsk in particular was discussed as a mistake of planning because of miscalculations around the actual availability of raw materials. The depletion of natural resources became one of the main problems of eastern colonization and a symbol of disappointment among specialists who had hoped to turn a new technological corner in Siberian and Far Eastern history.

Despite the fact that the Soviet government aimed to intensify the reforestation of "clear-cut zones of economic importance with valuable species," the actual share of reforestation remained small.[71] By the end of the Soviet epoch, specialists saw the negative effects of intensive wood harvesting in both old and new forest industrial regions. The disappointment rested with the fact that the gap between cuttings in old and new regions did not change substantively in the structures of wood consumption. Instead, the colonizers had explored and, as some put it, devastated forests located near transport infrastructures and most accessible territories.[72] By the mid-1980s, due to intensive cutting in the region of Buryatia, the River Tabur became three times shallower, and "after the total clear-cutting it became a thin stream." Hundreds of thousands of hectares of forest perished, the hunting sector suffered from significant losses, and air and water basins became contaminated. As one specialist put it, "The scale of the problem is huge and in some regions, because of their exploration [*osvoenie*], it is getting even worse."[73]

In spite of well-developed aerial photography in the advancement to the eastern areas, just under half of all forests were monitored by the end of the Soviet epoch.[74] This situation echoed the trajectory of industrial forests in the European region; by the late 1980s, because of clear-cuttings on the Kola Peninsula, air pollution caused by nickel production crossed political borders and became a serious problem for Finnish Lapland.[75] In some regions of Karelia, enterprises illegally clear-cut forests despite the fact that it was forbidden by state legislation.[76] Specialists complained about the negative environmental effects of this type of cutting, which decreased the water protection function of forests and productivity of forest soils. They indicated that foresters cut all the trees—even those that hardly reached ten centimeters in diameter—and described this as a loss for the national economy. Destroying young trees hindered future economic needs for wood.

In general, hopes that the imperial solution would solve the wood crisis were met with disappointing results early on, even in the decade after wood exploration in the east had unfolded. Soviet industry could not afford expensive infrastructure and continued with existing wasteful practices in the new lands. It was even more problematic due to the change in the old regions when, during reforestation, species changed; after they were cut, more industrially valuable coniferous species were naturally substituted with much less valuable broadleaf species. Trees planted by humans were more vulnerable to insects—a point of concern despite the fact that some commentators argued that it was not "a large-scale change."[77]

There was a tension between the availability of forests and value of wood. Huge green lands did not translate into a ready and plentiful supply of wood, as not all wood was industrially useful.[78] The relocation of cuttings to Siberia and the Far East implied that a new system of forest fire and beetle protection should be developed there, but it simultaneously required more funding and administrative efforts, which were both lacking in the Soviet economy.[79] Relocation to the east, which to a large extent remained in the realm of dreams and a point of utopian hope, required not only expensive labor but also serious technological efforts and costs to build equipment as well as technological infrastructures to harvest and process wood. By the late Soviet decades, the disappointment that arose concurrent with ongoing colonization was shared among intellectuals. In February 1977, Soviet writer O. V. Volkov wrote directly to Brezhnev, warning that the situation in most forested regions was critical. Calling the landscapes he saw with his own eyes "sorrowful pictures," he provided numerous examples of what he called an "irrational" approach to forests. He described them as follows: "I was once in deserted pine forests of Altai and saw mountain rivers filled with dry and abandoned logs . . . bald banks of River Sukhona and shallow banks of Selenga, walked around sawn away trunks in taiga in Primorye." Resonant with what later ecocide historians would take as their main argument, Volkov emphasized that "there are empty spaces everywhere," and "harvesters cut only the most valuable and leave a lot of incomplete felling," which wasted forests.[80]

Another problem in the east lay in the quality of the natural resources discovered. Most Soviet enterprises consumed coniferous wood, which they considered the most valuable in terms of industrial productivity because of the higher-quality cellulose that could be extracted from it. Due to its industrial value, this type of wood was most in demand by Soviet industry. The most widespread tree in the eastern regions was larch, which covered more than 60 percent of Siberia and the Far East; pine trees covered slightly less than 20 percent, while the most industrially valuable material, the fir tree, covered only 8 percent.[81] What specialists revealed was that the Siberian pine tree was less productive than the European one as it contained less cellulose.[82] At the same time, Soviet enterprises were technologically capable of using just 9 percent of each larch tree, while US producers used 30 percent and the Japanese used 58 percent. In fact, there was an oversupply of coniferous forest in the Soviet Union, pointing to the constrained ability of Soviet industry to process wood of lower quality.[83] Some specialists stressed that eastern forests were less useful in terms of accessibility and productivity, and thus in the European and Ural parts of the country, "there were the most valuable wood stocks."[84] In addition, a large part of the eastern lands was covered with aspen, which was not appropriate for industrial production and was less productive than coniferous wood from the Soviet European region. What appears on maps as the green land of the east was, then, from a purely industrial point of view, not actually rich in usable resources. This more pessimistic view of experts challenged the initial claims about the benefits of the relocation of wood harvesting in decreasing the industrial burden on forests.

Ultimately, the imagined treasures of the east proved economically and technologically less profitable sites, producing what this book calls "grand disappointment" in technological advancement into unspoiled green lands, even as the image of eastern abundance was kept alive outside the world of professionals. Some specialists underscored the tension between the image of plentiful forests and availability of industrial wood in the eastern regions, proving that both industrialists and alarmists had been mistaken to rely on the perceived richness of Soviet forests. In their emphasis on the need to save the disappearing forests of the European regions, both had relied on the notion of forest abundance in the east; both aimed to harvest more but save the wood stocks from total depletion. If industrialists continued to

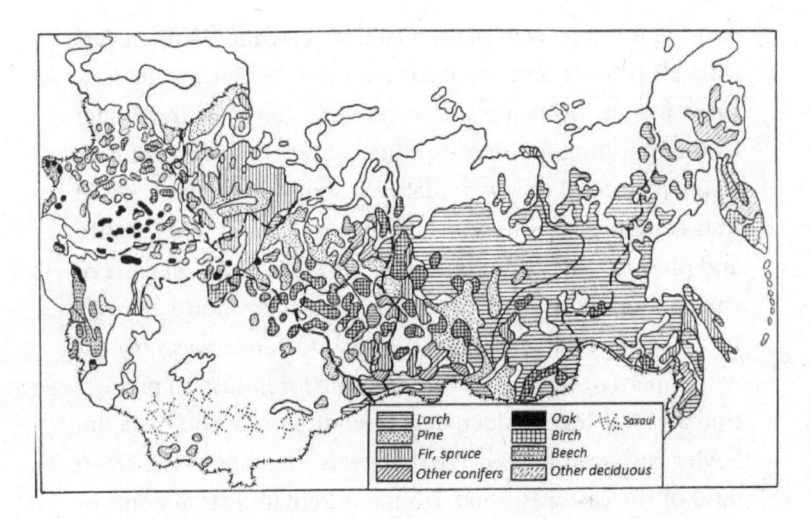

Figure 2.2 Distribution of tree species in the Soviet Union, mid-1960s. *Source:* Vasiliy Rubtsov and Prokopiy Vasil'ev, *Lesnoe khozyaistvo SSSR za 50 let (1917–1967)* (Moscow: Lesnaya promyshlennost', 1967), https://www.booksite.ru/fulltext/za5/let/2.htm.

insist on the richness of eastern resources, however, alarmists grew even more skeptical about the future of new regions. As one engineer put it, "We have to refrain from a vision of Buryatia [in Siberia] as a densely forested land"; in fact, he noted, most of its forest belts were located in remote and hardly accessible zones while the most accessible forest grew along the Transsib railroad. As a result, he argued, seemingly abundant forest resources left lespromkhozy "hungry" (*na golodnom paike*) because the distance required to transport wood to them was well over two hundred kilometers. Most accessible forests had been exhausted, prompting the degradation and sometimes liquidation of foresters' settlements.[85] According to E. I. Gorbatov, the head of one of the logging spots in the Irkutsk region, "The destiny of Siberian forests sent a warning. After wood harvesting, bare areas [*ogolennye prostranstva*] were left in places where no reforestation measures were taken."[86] Continuing to insist on the critical line, another specialist complained that in just fifteen to twenty years, the stocks of ash could be totally devastated.[87] The expectation of a wood crisis in the eastern regions remained strong at that time and translated into a general concern over the destiny of all the forests in the USSR.

Deriving from alarmism, forest colonization to the east was not initially aimed at clear-cutting but instead was an attempt to introduce a more productive dialogue between Soviet industry and nature. The aim of providing the industry with sustainable wood resources was underpinned by the desire to cut less in the European part and save natural resources from total depletion. The very idea of relocating main industrial cuttings to the east, even though it failed technologically, demonstrated how industrial ecology was evolving in the late Soviet Union. This dream met the harsh realities of the natural conditions of eastern forests and was further obstructed by technological weaknesses of the planned economy, which did not supply enough technological resources for the scale of the ambitious projects. As such, the colonization project miscalculated the material possibilities of state socialism as they were. Disappointment about the failure of technologically equipped colonization and the difficulties of accessibility presented by the eastern forests intensified the experiment as well as the search for alternative resources to compensate for the lack of industrial wood.

3

THE QUEST FOR A NO-WASTE ECONOMY

THE RATIONAL AND COMPLEX USE OF NATURAL RESOURCES

Economizing through saving on the costs of production and resource use became a key model underpinning specialist explanations of the direction of Soviet industrial policy.[1] In particular, minimizing material expenses formed one of the key motivations in the workers' rationalization movement that developed from as early as the Stalinist industrialization era of the 1930s. The concept was revived and flourished once again in the mid-1950s, as it would once again do so in the later decades of the Soviet epoch. In industry, rationalizers were workers and engineers who came up with useful professional ideas as well as proposed savings on the costs of production through partial mechanization and small improvements in technological processes invented on-site. Economizing became part of industrial rhetoric and a crucial measurement for industrial productivity. In forestry, it had two connected implications. On the one hand, the industrialist approach developed methods to save resources at enterprises against the backdrop of the shortage economy. The main method entailed saving on material resources for producing and reducing wastage, thus overcoming frequent stoppages that arose because of material shortages. On the other hand, economizing had an environmental dimension: for many forestry specialists, it seemed to offer a way to prevent the rapid depletion of wood resources.

Economizing justified the use of alternative raw materials as substitutes for wood to make production cheaper. Using alternative resources, such as various types of industrial and consumer waste, promised to help save costs related to extracting new materials and therefore increase the effectiveness of the planned production. First, the approach emphasized the role of wastepaper (i.e., the by-products of consumption). Second, it pushed specialists to reconsider the role of industrial waste materials left by wood harvesting and sawmilling (i.e., materials extracted from wood and left as waste in forests and at industrial enterprises). Third, it enabled specialists to propose various projects for using annual plants (such as reed) and agricultural by-products (like straw) as alternatives to wood. Specialists explained that economizing could be reached with the implementation of the so-called complex use of natural resources—a concept rooted in the pre–Second World War Soviet economy that emerged out of the economizing imperative stressed in Soviet industrialization, which faced numerous shortages of supply and at the same time the pressure of the plan.[2] Thus a 1932 brochure issued by the Scientific-Technical Council of Paper Industry on raw materials argued that the increased demand for softwood meant that *all the parts* of cut trees should be used instead of only the most valuable portions of wood.[3] This brochure highlighted the unique ability of the socialist regime to pursue techniques of economizing and complex use, insisting that the Soviet planned economy, industrialization, and collectivization helped save materials, while by contrast, capitalism was inherently and irredeemably wasteful. Many enterprises practiced economizing, prescribing that workers save resources, including raw wood, chemicals, and electricity. In Soviet industrial discourse, the invocation of complex use was explained by the drive for exploiting natural resources while avoiding the wastage of resources, understood as industrial value that could be salvaged through new technological possibilities. All parts of the tree were to be used, "from the roots to the top of the tree," along with all the waste received after processing wood through sawmilling and other industrial operations. These salvaged materials could be used for manufacturing ready-to-consume products. Complexity, then, was a method for achieving cost savings. As early as the immediate postwar years, Soviet forestry specialists aimed to work out no-waste technologies that originally derived from an industrial discourse, proposing that the industry find new sources of raw materials to satisfy the expected rise in consumption levels.

Reviving earlier professional discussions of complexity, engineers and industrial scientists in the 1950s insisted that it was important to use wood effectively, connecting the complex use of resources to rationality (*ratsional'nost'*) and both to modernity. Leaving potentially useful industrial material as waste was now explained as an irrational, wasteful, and backward practice. Rational decision-making was, by contrast, to be achieved through precise calculation, which in turn emphasized a no-waste approach—because all waste could potentially be transformed into economic value in modern technological production. Environmental rationality was framed as a form of socialist modernity; the actions of the modern (*sovremennyi*) individual should be guided by calculations of practical value and the need to consume natural resources in a more sustainable form. Late Soviet rationality as a modernist practice appealed to finding tools to increase the level of the sustainable consumption of resources. Rationality was an antidote to resource depletion; as one specialist put it, "If we have the *right attitude* toward forests and locate main logging capacities properly, our forests will never be depleted."[4] Science, technology, and a "proper attitude" were thus important prerequisites for sustainable wood supply and preventing the wastage of forests as valuable industrial resources. *Struggling*—the term often used by Soviet specialists in the spirit of the epoch—to economize each cubic meter of wood was a crucial task, the effective means by which costs of raw materials would be reduced.[5]

The intertwined discourses of complexity and rationality used by forestry specialists emerged as a response to the expected wood crisis. First, specialists suggested that modern technology could transform waste from rubbish into used value if guided by expert knowledge and skills, and employing technological infrastructures. Second, these discourses grew out of the chronic shortage of raw materials and evolved as a solution for deficiencies in resource supplies. Third, their purchase grew against the backdrop of escalating disappointment in Siberia and the Far East discussed in chapter 2. While deriving from the internal evolution of Soviet industrial ecology, new thinking about the relations between industry and nature among specialists echoed a growing environmentalism in Western countries. Rising voices against the use of some chemicals in agriculture along with the application of atomic science and radiation did not find many supporters in the USSR, but some Soviet specialists shared a general vision of the need to move toward a greener economy to reach economic effectiveness. The anticipated depletion of industrial natural resources due to the rapid deforestation of available forests, combined with the problems of exploring the distant and difficult-to-access forests of the Soviet eastern regions, presented significant motivation for introducing complexity and rationality of resource use there too. Overall, this led to the adoption of a paradigm that combined the discourses of the complex and rational use of natural resources as the main approach to industrial production. While the twentieth meeting of the Communist Party held in 1956 was renowned for the launch of the de-Stalinization program, it was also here that the leadership declared the imperative to develop the rational use of raw materials, fuel, electricity, and other resources. Speakers at the meeting maintained that the complex use of resources, diversifying the range of products made from waste, and decreasing waste losses were all required for economic development. Many criticized Soviet producers, asserting that they had not worked to save national riches and the struggle to economize material resources had to become the subject of everyday concern for every worker, engineer and producer alike.[6] As ministry officials discussed further in the 1960s, in light of earlier intensive clear-cuttings as well as the increasing distance between wood-harvesting spots and wood-processing enterprises, it was important to use all the available resources in situ instead of relying on complicated logistics in wood supply.[7]

Since at least the 1960s, many forestry specialists insisted that the combination of the complex use of wood along with a high degree of mechanization and automation would spearhead the manufacturing of diverse products from wood, consumer and wood waste, and annual plants. As a result, it was hoped that growing economic and consumer needs could be satisfied by means other than increasing wood harvesting—through the more efficient use of wood and use of alternative resources in industry.[8] As one ministry specialist stated in 1973, "The organization of full and rational use of forest resources based on continuous forest use and timely reforestation, as well as planned satisfaction of economic demand for wood, are the main problems of forestry and the forestry industry." He continued that "notwithstanding how large our forest riches are, we cannot *simply increase* the volumes of cutting."[9] Rationality and complexity therefore proposed an alternative to the colonial solution to the wood crisis, representing what can be called the productivist approach to industrial forests.

This idea further intensified from the 1970s on, when economizing and the eradication of resource losses became even more pronounced discourses in wood harvesting and the manufacturing of wood-based products. As with the Stalinist industrial discourse of complexity, by the mid- to late 1970s, rationality and economizing were declared as a "Communist feature" (*kommunisticheskaya cherta*) and "the norm of our life," urgent and necessary.[10] Moreover, Soviet professional publications declared that the saving of material resources was an element inherent to the very nature of Soviet socialist planned economy.[11] In 1970, the Ministry of Forestry and Wood Processing Industry issued a decree that obliged the heads of industrial enterprises and logging spots to "liquidate the irrational use of wood." The economy was centered around saving all resources; total saving allowed for the production of more consumer goods despite the shortage of resources.[12] The need to "struggle for rationality" was declared and frequently repeated by employers of various branches of the forestry industry in their professional journal publications and industrial reports. Economizing implied preventing wastage of *seemingly* unneeded resources. In this sense, specialists saw wastage as a temporary problem—a capitalist hangover in the Soviet economic system that obstructed the goal of outpacing the West in the economic competition.[13]

As some specialists wrote in 1979, the Soviet economy had consumed the "best wood," but disregarded the potential value of the rest of it.[14] They declared this attitude to be backward, contradicting the modern principles of the economy and industrial production. Specialists sometimes labeled the habit of leaving waste in the forest as old, inefficient, and even barbaric, proclaiming it should be rejected as a sign of backward times. It implied producing more material goods and consuming fewer resources as two imperatives that moved the industry and particularly wood-harvesting sector. Consumer and harvest waste, along with other alternative resources such as annual plants, came to be viewed as materials that could provide the industry with resources and potentially decrease forest cutting. Specialists believed that complexity and rationality could help the industry solve the deficiencies of the extensive model, compensating for the lack of the most valuable and devastated coniferous wood. Wood-processing enterprises usually tried to refuse supplies of deciduous wood, denigrating it as a less productive type from a technological perspective.[15] Complexity, however, implied that enterprises could manufacture technologically simpler products, such as cardboard, from industrially illiquid wood, which was in demand for mass consumption, but was quite literally thrown away at logging spots and enterprises. Specialists asked why it was necessary to use high-quality wood to make low-quality packaging paper when one could use consumer waste and wood waste left in forests instead. Growing consumerism indicated a growing demand for simpler-to-produce goods, ranging from napkins to wooden toilet seats, which in fact did not require high-quality wood to manufacture.

From the industrialists' perspective, nature served the industry in terms of offering its resources, which had to be used rationally to sustain their availability for a long time.[16] Rationality as an instrument of economic policy was intended to help treat nature, the "green pot," carefully and not deplete it but rather unlock its utmost potential to provide economic benefit.[17] Overall, the drive to use alternative materials for manufacturing in place of wood was conceptualized as an important means to produce more resources at a lower cost. Embedding the use of wood alternatives as an urgent economic goal, the nascent environmentalist concerns of specialists and the state obviously remained constrained by industrial discourse. Saving natural resources was crucial for making more products in the future, and using various additional materials served this task too. Many specialists argued that the enormous amounts of wood waste left in forests as well as at sawmills and wood-processing enterprises along with the paper waste left after consumption (albeit in much smaller of amounts) should be reconsidered not as waste but instead as potential alternatives for wood that could provide a sustainable raw material base for industry.

WASTE AS A MATERIAL OF MODERNITY

Following the paradigm of rationality, specialists came to conceive of alternative raw resources—the term that this book uses for describing different types of waste and annual plants—as modern materials in industrial production. The common definition of the concept of *waste* denotes that which is not appropriate for any further use, left behind and incapable of satisfying users' needs, and not needed.[18] After the Second World War, the Soviet forestry industry witnessed a transition in thinking around many materials due to a new economic model of resource use that developed among experts: from waste to resource. Waste materials were reconsidered as valuable resources that could be used in industrial operations in addition—or even entirely instead of—critically disappearing wood. Over time, alarmism increased even further, entrenching expert's conception of waste as, in Marxian terms, "use value."[19] Industrial (harvesting and sawmilling) and consumer (used paper, sacks, etc.) waste was considered "a hard burden" for the economy if it was not used.[20] Some specialists criticized how, *in the past*, Soviet industry had seen alternatives as useless and, as some put it, "evil." In 1964, for example, about 43 percent of each cubic meter of harvested wood was considered waste, meaning that a little over half of it was manufactured into wood boards, sleepers, packages, and so on. Some calculated that the use of waste industrially would increase this proportion to 90 percent.[21] Through the use of waste, it was believed, economic materials could be produced with less intensive exploitation of natural resources.

Waste, specialists said, should be transformed into economically useful material through the power of modern technology and experimental efforts, and specialists were to give greater regard to them. Engineers and scientists who worked at industrial enterprises and research institutions conceptualized these industrial waste materials as magical resources that would solve the problem of deforestation amid the voracious timber demands of enterprises. Wasting was also connected to inefficiency; from the perspective of specialists, it was not efficient to leave large amounts of potentially useful resources as waste. Alternative raw materials were plentiful, available, and more sustainable than cut trees. Raising the share of wood waste and other alternative resources in the production of consumer products increased the degree of rationality in resource use, according perfectly with the discourse of economizing.

Growing optimism over the industrial power of waste found its support in modern science and technology. At the close of the 1950s, chemistry promised change on a global scale, bringing far-reaching advances in people's nutrition and the quality of their material life. Now chemistry entered the forestry industry to offer crucial tools for achieving economic prosperity and abundance. Specialists saw forestry chemistry as the means of achieving "a radical solution to the problem of [the] rational use of fire and low-quality wood, harvesting, sawmilling, and wood-processing wastes."[22] Chemistry could thus transform waste into ready goods, rendering wood a material of modern and rationally functioning industry. Discourse around the transformative ability of wood emphasized the binary between, on the one hand, the traditional and, as many put it, primitive usage of wood, and on the other hand, modern or perceived progressive industrial methods. Technological achievements espoused by specialists assured the Soviet regime that it was becoming increasingly possible to build "the material basis of Communism" quickly and increase consumption levels in Soviet society—a pivotal issue in facilitating modernization in everyday life. Wood was a vital material of modernity, the importance of which was increased because of the Cold War rivalry to build a more modern society through increasing the manufacture of mass consumer goods. As such, the extraction of natural resources "in ever larger quantities" was not only, as some researchers have suggested, "the means of supporting the needs of the country's military as well as its energy-intensive economy" but also central in driving the shift toward consumerism.[23]

Specialists saw technology as a means of transforming waste material into modern products that could serve as a substitute for wood in traditional products (such as writing paper) and new goods (such as cardboard packages). Many wrote, for example, that the magic of chemicals could transform tree leaves into vitaminized powder and essential oils for the food industry. Soviet research and industrial reports explained that producing new types of plastic from compressed wood waste chips was crucial to replace "traditional" materials, such as expensive aluminum and copper that were in short supply. As engineers wrote, "[We] rely on the possibility of using wood plastics and [seek to] considerably increase their production for details in machinery making instead of metal." Rapidity was important here to introduce new and cheap modern materials. As some specialists argued, "The national economy would receive enormous savings on nonferrous metals and other deficit materials."[24] Previously, they explained, this innovation was delayed due to the "low quality and difficulty of the technological process," revealing industry's lack of resources for implementing this progressive process.[25] In the 1950s, other applications of this process became particularly crucial: wood plastics were proposed for use in the manufacture of chipboards for construction and housing materials.

Another application of wood waste was furniture, especially kitchen cabinets, tables, wardrobes, and beds, which was in demand for the new individual apartments called *khrushchevkas*, built by the state as a solution to the housing shortage. Since the 1950s, most furniture was made from various modifications of so-called high-density fiberboards manufactured from pressed and laminated wood-wool, a waste product from sawmilling and wood processing. In 1962, up to 80 percent of panel furniture and other house details, such as doors, sidings, and floors, were made from fiberboards.[26] Soviet scientists revealed various characteristics of these fiberboards: unlike many wood-made products, they did not dry out, lose their shape, or burst, and were easily decorated. Making fiberboard furniture was quicker and easier industrially. Beginning in the mid- to late 1950s, as part of the postwar program to solve the problem of housing shortages, thousands of new residential buildings were introduced in the USSR. Industrialists and planners stressed that it was important to provide people with modern furniture that was both beautiful and cheap. They also suggested that consumers themselves could assemble furniture at home, thereby saving on manual operations in industry. This concept was similar to that seen in contemporary assemble-it-yourself furniture companies like the internationally renowned IKEA. In 1985, the Project and Technological Institute of Furniture of the Ministry of Forestry, Pulp and Paper and Wood Processing Industry of the USSR developed a number of initiatives for making furniture in a series titled *Do It Yourself*. The consumer could choose elements to make up a furniture item, which they could then complete at home. As industrial designers explained, consumers became "coauthors" of professional designers and so expressed their own creativity in materiality.[27]

Figure 3.1 Many Soviet publications advertised synthetic materials extracted from cellulose. *Source:* Anatoliy Averbukh and Kseniya Bogushevskaya, *Chto delaet khimiya iz drevesiny* (Moscow: Lesnaya promyshlennost', 1970).

Some specialists also proposed using wood waste as biofuel. Compared to simply burning wood waste, a practice heavily criticized by many specialists as a waste of valuable materials, the biofuel extracted from wood wastes could economize the costs of production and hence was an environmentally friendly modern practice.[28] Others criticized the enterprises because they did not utilize tree bark, which was mostly discarded as rubbish. Burning bark was declared dangerous for the environment, and some specialists claimed that it would be better to use bark for fertilizing soils.[29] In the 1970s, waste was also used for manufacturing souvenirs, such as *matryoshkas*, Santa Clauses, and other small goods.[30] In many cases, it would help shift the production of consumer goods from handmade to industrial scales. Specialists stressed that waste was a substituting material; a cubic meter of chemically processed pine tree wood could, for example, provide 160 kilograms of synthetic silk or 170 kilograms of wool. By contrast, "direct" producers could offer significantly less: a silkworm produced half a kilogram of silk while "the best sheep gives 6 to 7 kilos of silk per year."[31] In production, as specialist E. Lopukhov estimated, making ready products from waste would make them up to 20 to 30 percent cheaper than their previous cost.[32] Another engineer argued that by the year 1970, according to his calculations, the need for wood would exceed 850 billion cubic meters. It was an incredible number, which the Soviet wood-harvesting industry was simply incapable of supplying. "This is why," he said, "we will destroy all the economic plans if science did not find new ways [to use waste material], because waste is *more valuable* than wood."[33] A lack of action in this sense was seen as a potential economic loss, intensifying fears about the future.

Modern technology processed natural materials as well as helped the industry solve the contradiction between the lack of industrial forest resources and consumer demand. Using alternative materials in industrial production was an experimental solution to this dilemma and attempt to develop more sustainable manufacturing. The variety of waste in forestry was wide, and an increasing number of items came to be considered as valuable resources. Waste thus emerged as part of industrially embedded ecology as the implied "savior" of industrial forests, moved by the economic purposes of modernity. The saving effect of waste was often emphasized in late Soviet propaganda films, which drew a connection between no-waste production/consumption and a relationship of care to the motherland.[34] Here, discourses of personal duty, social responsibility, and the waste economy overlapped.

CONSUMER WASTE GAINS INDUSTRIAL VALUE

The 1950s were particularly important for reconceptualizing paper waste. Several professional initiatives investigated ways of using paper waste in industrial production aimed at sustainable economic growth. Despite the fact that consumption levels of paper were fairly low, wastepaper was believed to provide a solution in the search for alternative resources in the forestry industry. In June 1955, the Academy of Sciences of the USSR held an important meeting on the use of waste where officials of high rank gathered, including the influential minister at the Ministry of the Forestry Industry of the USSR and "patriarch of the Russian forest" Georgiy Orlov, minister of the Paper and Wood Processing Industry

Feodor Varaksin, and well-known academicians such as forest specialist Vladimir Sukachev.[35] Their discussions resulted in the conclusion that new technology and forms of industrial organization for the use of waste were urgently required.[36] In realizing this task, they understood waste in a broad sense, stating that everything ranging from wood harvesting, wood processing, and sawmilling to consumer paper and paper-based products constituted usable waste products. Urgency, as it was formulated at the meeting, reflected the fact that the participants saw modernity as a challenge and held that measures to use alternative raw materials constituted a modern practice of production as well as consumption.

In 1956, the Central Institute of Cellulose and Paper Industry, the only institute of its kind in the country, held another meeting of its economy section, during which specialist Feodor Kuteinikov reported on the use of wastepaper in industrial manufacturing. Trained in economics, Kuteinikov worked in the forestry industry and devoted many years of research to the use of wood, paper waste, and other materials in industry. In his presentation, he explained that "the growth of industrial and individual consumption increases the amounts of industrial and household raw materials, which can be a solid basis of our socialist industry. The rational use of waste materials acquires quite an important economic meaning because it enables the increase of amounts of industrial raw materials." Kuteinikov estimated that about one ton of wastepaper could replace one ton of timber. If his calculations were correct, however, the collection of wastepaper and cardboard in the Soviet Union (16 percent) was half that of Western countries (30–35 percent). The reason for this was the lower levels of consumption of paper and cardboard, on the one hand, and the backward infrastructure of waste supply, on the other hand. In addition, it was a lack of knowledge among Soviet consumers about sorting and preparing waste for recycling. He provided the following example: consumers submitted paper waste without preparing it properly (not removing paper fasteners and other non-paper elements), which required more workers to prepare the waste in the absence of any automatic cleaning or sorting systems for wastepaper in the USSR at that time.[37] This is why a significant portion of collected paper was thrown away as waste—that is to say, waste from waste, because only the cleanest part of it could be used for industrial production. While this recursive approach conceptualized waste as a service for industrial manufacturing, it largely depended on consumer action.

Kuteinikov gave another illustration: a rubber-making factory supplied soot sacks to the Leningrad factory of technical paper for recycling but did not clean them. The paper factory refused to recycle them because the whole process of cleaning the paper required five additional days of work in what were described as unsanitary conditions. As a result, the rubber-making factory burned three tons of waste just fourteen kilometers from Leningrad, the second most prominent Soviet city. He also stressed that workers sorting waste were not well qualified and well versed in the technology of papermaking and cardboard making for which the waste was collected. They did not understand that different forms of waste were used for making different types of paper and cardboard. Kuteinikov complained, "We have been projecting recycling enterprises since the 1930s, but until now we have not constructed anything. Ministry workers did not come to waste disposals to see how much valuable raw material is just *dying* there."[38] For the years of Soviet power, he continued, "we have not constructed any waste recycling enterprise and wastes are recycled at enterprises which have historically developed conditions for recycling facilities."[39] Enterprises did not take into account that it was important not only to collect wastepaper but also to invest in the infrastructure needed for preparing waste for recycling. From his point of view, then, the problem of consumer paper waste in the Soviet Union lay in weak knowledge about the specifications of industrial operations that were seen as simpler than they were in reality. Animating waste as a living organism left to die and perish in ways that were similar to how forests were often described, Kuteinikov revealed a weak waste consumer culture, including the sorting and cleaning of waste, at the individual level. As specialist Viktor Mudrik put it in 1971, "We have an enormous gap between the possibilities and demands."[40] Some workers and engineers complained that the wastepaper they received was decaying and not usable for making goods of high quality.[41] This spoke about a gap between producers and consumers, or those who delivered and used consumer paper.

Referring to the West as an example of desired practices, specialists frequently criticized their domestic model of waste use. Thus as one said in 1956, in the United States it was forbidden to wrap food in newspaper, even a new ones, but in the Soviet Union, *used newspaper* was delivered to factories that supplied paper to the Ministry of Food Industry—and sometimes they sent wastepaper to food enterprises directly (he did not clarify the purpose of it, though). As with other components of the Soviet–United States Cold War rivalry, specialists saw a US waste culture as more advanced, modern, and civilized. This was the counterpoint that spurred a general drive in the Khrushchev era to increase the quality and hygiene standards of production and food consumption. As the discussion of 1956 showed, engineers imagined that countries of the West more generally were far more advanced in the recycling of waste. As was noted by female specialist G. I. Shcherbina, Swedish enterprises sorted forty types of paper wastes (compared to thirteen types in the USSR) because of "excellent technology" and a *proper* waste culture.[42] Soviet waste-recycling technology to a large extent remained a matter of professional imagination rather than a real practice, and many specialists criticized this fact.

Overall, many specialists were critical of the waste problem in the USSR, connecting wasting, sustainable industrial development, and consumer culture. Discussing waste culture in the Soviet Union, specialist A. Shuko asked, "Tell me, how many [old] newspapers in your life you have submitted for recycling? I have submitted nothing, and I think that you all are similar to me."[43] He also stressed that the question of recycling had been discussed since the early 1930s, but had not led to any fruitful results because of the lack of infrastructure.

Yet even as they continued to emphasize the high need for using waste, later specialists complained about a poorly developed wastepaper culture and blamed infrastructural obstacles. In 1971, at the conference held in Leningrad on new technologies and equipment for recycling used paper, specialists mentioned that industrial enterprises practiced fraud and delivered waste of mixed quality, placing good-quality waste above a mass of far worse quality. One also criticized the fact that specialists had talked about the same problems for decades but with no effective result: "If you listened to conference papers yesterday, you might note that we [still] discuss the same issues we did forty to fifty years ago, including such basic things as the sorting and cleaning of waste." In fact, many complained about how waste materials supplied from industrial enterprises were sometimes sullied with mud, food, and plastic waste, which littered filters when recycling. As the same speaker underscored, "We receive waste with so many additional components that even workers are afraid of getting tons of bad-quality paper [produced from waste]."[44]

On the experimental level and in experts' imagination, the use of waste was successful and the appeal for the use of consumer waste grew stronger by the end of the Soviet epoch. The idea of saving raw materials and making no-waste production seemed to have reached the peak of its popularity by then. In 1982, for example, engineer D. Kovaleva wrote enthusiastically that each gram of wood waste was valuable for the economy because "wood is forest that required decades to grow, [and] a lot of money and labor to invest."[45] While consumer waste remained smaller than wood and sawmill waste, the latter became the subject of experts' concern even more often. Overall, however, the recycling of wastepaper in the USSR grew but did not reach sufficiently high levels; it produced 826 thousand tons in 1970, 1,050 thousand tons in 1975, 1,300 thousand tons in 1980, 1,533 thousand tons in 1985, and 1,590 thousand tons in 1988.[46] Recycling thus did not increase radically: while in twenty years it almost doubled, it was still much lower than in many other countries.

The problem lay in the fact that the volumes of waste use were not sufficient while at the same time supplies of wood could not meet demands for making pulp and paper-based products. Viktor Mudrik, a specialist from the Moscow headquarters of Giprobum, the planning institution for paper industry, stressed "the enormous gap between the possibilities and demands."[47] He complained that even the biggest factories could not fully use wastepaper, emphasizing the problem of recycling materials. Experts' complaints therefore revealed the lack of technical possibilities despite the spread of more progressive thinking about industry-nature relations and consumption culture.

WOOD AND SAWMILL WASTE ARE PUT TO ACTION

From the mid-1950s on, some specialists of the sawmilling industry discussed not only the industrial importance of paper waste. They also raised alarm about the huge amounts of tree components (such as branches, bark, and roots) that were left as waste in forests, rendering them 'littered' (*zakhlamlennye*), and at industrial enterprises. They viewed this abandoned material in forests and at enterprises as both a waste of crucial industrial materials and exacerbating the wood crisis. They particularly noted that the total sawmill waste made up a large share (36 percent) of all wood in the sawmilling industry.[48] The search for better usage of harvesting and sawmilling waste arose together with concern around consumer waste, spurred by alarmist warnings of the finite availability of resources. Now specialists recognized that developing technologies for processing these waste products was needed to help increase productivity in the use of forest resources. Referring to forests as a national treasure and "green pantry," many specialists also spoke about wood and sawmill waste as a potential green pot.

The use of wood and sawmill waste in this sense implied the possibility of better use of raw materials (from low-quality wood and various parts of cut trees) in manufacturing demanded products. It would also help leave forests clean and decrease the number of forest fires—a problem that was among the biggest in the Soviet Union, as it remains in modern Russia and other countries (one might recall the recent devastating fires in Australia, the United States, and Greece).[49] Using waste and, as sources often put it, *struggling* for each cubic meter of wood meant discovering (*vskryvat'*) how the riches of resources could be made rational in industrial operations, decreasing the costs of energy needed for production along with the amount of wood used.[50]

Primarily, the logic of rationality and complexity as well as the imperative to decrease quantities of "wasted waste" evolved from industrial priorities. In 1970, the amount of forestry waste produced in the USSR neared 220 million cubic meters per year, and still only 15 percent of this was used as raw materials while the rest was burned as fuel at enterprises.[51] In 1973, the increase in the amount of waste used by the forestry industry (from 11.2 million cubic meters actually used in 1970 to 23 million to be used by 1975) was included in the prospective plan of development of forestry, wood processing, and the pulp and papermaking industry between 1976 and 1990.[52] Some enterprises indeed successfully developed the use of waste material, such as the Beregometsk forest combine in Ukraine, an unforested region that traditionally suffered from a lack of wood; by 1979, over

60 percent of the raw materials they used was constituted of recycled waste.[53] The Kostroma plywood plant was another story of success. In the mid-1970s, the plant used up to 80 percent of all the waste it produced for manufacturing fiberboard, largely used in the furniture industry. It also manufactured coat hangers and ice cream sticks from dry veneer using machinery constructed by the plant's inventors, and supplied these goods to shops. Through developing the technology of mixing cut plywood waste with synthetic glues pressed at high temperatures, the plant made toilet seats and stools, which were in short supply in the USSR.[54]

Despite successful examples, the share of waste use in industry remained at just 10 percent by the end of the Soviet project.[55] Even at this point, almost half of the harvested wood was still left in forests while the enterprises used less than 10 percent of wood and sawmilling waste material. Providing concrete calculations, specialists said that enterprises usually utilized just 8 percent of waste but frequently burned it as fuel as they had done before. "Other waste (41 percent) just *dies*, 16 percent decays in the forest, and 10 percent is burned while 15 percent just lies in the store houses."[56] From a hundred harvested cubic meters of wood, only half was delivered to consumers, while waste made up almost 50 percent of the harvested wood. A considerable quantity of waste was transported to a disposal area, or in professional slang, *v otval*. This happened not only in the old forested regions but in the new eastern lands too. Hence the Maklakovo-Yeniseysk region in Siberia, chosen as the most appropriate for the complex use of wood by Soviet colonizers, consumed only a limited amount of waste. As the Siberian branch of the Academy of Sciences reported to Moscow, "Wood waste does not find its full use and is thrown away or just burned," representing big losses for the economy.[57] According to other calculations, enterprises in the Far East produced more than four million cubic meters of wood waste per year but only used a small proportion of it: 11 percent was used for technical purposes, 1 percent for making consumer goods, and 14 percent was burned as fuel because "of the deficit of fuel resources such as coal, peat, and mudstone."[58]

Specialists complained that the progressive idea about using waste in numerous industrial operations did not have any infrastructural support, even in the new industrial forestry regions. Nor did enterprises have appropriate techniques and facilities for using waste industrially. If low levels in the use of paper waste were to a large extent connected to a poorly developed recycling culture among consumers, specialists believed infrastructural problems at enterprises were to blame for the low levels of wood and sawmill waste use. Indeed, most enterprises, especially in traditional forestry regions, often saw these materials as secondary and time-consuming, while their infrastructure was often inappropriate for waste recycling. As such, forestry logging spots complained that after clear-cuttings, they received huge amounts of branches, which no one enterprise wanted to take stock of.[59] The successful use of waste was largely dependent on the ability of enterprises to organize a separate industrial shop to make consumer goods.

Using waste therefore remained more a matter of industrial discourse than practical reality—a point of strong criticism against Soviet wood harvesting and processing among many professionals. As an engineer stated, even in the 1980s at most enterprises, "waste is used at a low scale mainly as [primitive] fuel. No one calculates how much of it is used."[60] Industrialists often blamed research institutes for the lack of adequate research and recommendations for using waste resources in industry. For researchers, the most problematic question remained how to make machinery for cutting waste into wood chips, the most important preparatory step in waste use. In 1990, specialist N. V. Sinyaev criticized the use of wood wastes. He complained that they were not used in Karelia at all but still simply thrown away, or at best used as fuel for heating. According to him, in the whole country in 1990, only 26 percent of waste was used and then mainly as fuel.[61] Others complained that enterprises still burned waste even in the newly established "enterprises of the future" in the eastern regions, thereby continuing what was seen as a barbaric practice. As the head of the Institute of Forest and Wood of the Siberian branch of the Academy of Sciences wrote in a newspaper article titled "Fires of Poor Management" (*Kostry beskhozyaistvennosti*), enterprises in the Krasnoyarsk region annually burned millions of tons of wood waste that could have been used in pulp and papermaking industries.[62]

Industry failed to realize plans to manufacture products like plywood, fiber wood plates, and wooden houses from waste. Due to the increasing diversification of paper products, Soviet consumer industry suffered from a lack of new products like receipt tapes with thermo adhesive lines and label paper. In the mid-1970s, the Gosplan, the main Soviet planning institute, insisted that the manufacturing of these two products "is completely insufficient."[63] Similarly, the industry did not produce enough goods that combined paper and polymers—products with "polymeric vaccination" as one specialist put it.[64] In 1976, the Department of Forestry along with the pulp, papermaking, and wood-processing industries of the Gosplan, concluded that the availability of forest resources was insufficient for future industrial purposes and constituted the reason "it is impossible to satisfy the needs of the economy" even for "the minimal needs." As such, the department asked the Gosplan not to decrease the rate of cutting in the European and Ural parts of Russia. Significantly, even despite the crisis in supplies with raw materials, Soviet authorities continued exporting timber abroad.[65]

The story of waste use entails two important observations. On the level of implementation, the use of various waste met infrastructural difficulties perceived by specialists as technological backwardness and the lack of needed facilities. This infrastructural deficiency of the state socialist economy was an obvious and tragic obstacle in the way of ambitious projects that could lead to the green economy. On the discursive and experimental levels, the use of waste was part of intensive discussions exposing "progressive" ideas about complexity and rationality as crucial conditions for modern production and consumption. This was evidenced in growing industrial ecology as a new industrial consensus, though restricted by technical possibilities and low investments. To a large extent, industrial ecology embedded in progressivism and economization remained an imagined and desired project implemented in only a few enterprises.

THE LIMINALITY OF FOREST

While industrially embedded ecology remained a discourse shaped by industrial priorities and aimed at production, it gradually saw the rise of more environmentalist attitudes among specialists. The use of waste was primarily connected to notions of economic efficiency and cost saving. By the late 1970s and early 1980s, a new line was gradually developing within the Soviet forestry industry. Some specialists increasingly spoke about the recreational function of forests as part of their economic meaning, arguing, as one observed, that "forest use is a multisided process" that requires "the rational use and reproduction of forest resources because forests are not only a unique part of natural environment but also an inherited part of [the] ecological [*ekologicheskogo*] and socioeconomic welfare of humans."[66] Sustainable forest use in this sense was still explained through a consumerist vision of nature, which implied that trees were not simply a source of wood but fulfilled recreational and aesthetic functions too. For these reasons, they were to be conserved for the present and future of humankind. This line emphasized the need to preserve forests. It did not mean, however, that forests were to be left untouched. Ideas about the sustainable use of resources were born in parallel with the gradually developing environmentalism along with criticism of the ecological externalities of modern technologies in the West, to a significant degree stimulated by Rachel Carson's *Silent Spring* (1962). In countries like Sweden, early attempts to develop greener production were undertaken around that time, seeking closer cooperation between producers, the state, and environmentalists. To a certain extent, the late Soviet Union's closer consideration of forests as a resource in danger was triggered by international discussions on environmental issues across the Iron Curtain.[67]

The notion of *sustainable forestry (neistoshchitel'noe lesopol'zovanie)* was rooted in the nineteenth century and well employed by Soviet specialists. In a 1964 article, A. Shcherbakov wrote that the use of forest resources would be possible in different areas for between twelve to fifty-five years, denoting the period during which forests could be cut while securing their bounties for the future. After that, he warned, a critical threshold would be crossed, initiating a kind of death spiral for forest resources. When asking the reader what would follow depletion, Shcherbakov stressed that no Soviet development plan had an answer. As he pointed out, wood harvesters knew well that trees cut yesterday would only regrow in fifty to sixty years. Only the future Communist society (which as the Soviet leadership declared would be largely reached by the year of 1980) could exploit these forest resources, Shcherbakov insisted; forestry employees should remember that the results of their activities would have long-term consequences and contribute to the prosperity of the future Communist society. Combining ideology and sustainability, then, he maintained it was important to search for new ways of "exploiting [the] forest riches of our fatherland," which could help produce high-quality but cheap wood as well as provide quick and firm reforestation in the coming years.[68] In his words, and as was indeed typical of Soviet discourse, it was urgent to use forests economically (*po-khozyaiski*) because, according to the ideological mantra, forests belonged to the Soviet people. Here he translated two ideas quite typical for late Soviet discourse around forest resources. First, he connected forests and natural riches, stressing that forests were an inherent part of national treasures, even though they were to be industrially exploited. Second, he discussed the regeneration of resources and their industrial exploitation as both efficient and quick. Conserving various resources was the main trigger for introducing complexity in the resource use, and the conversations over waste contributed to the sense of the increasing value put on forest.

Forestry scientist Nikolay Anuchin, who carved out a brilliant academic and administrative career under the Soviet regime, also contributed a great deal to the notion of sustainable forestry. In his explanation, he insisted that the volumes of a forest's use should not exceed its annual growth within a short period of time.[69] Similar to Shcherbakov's view of forest sustainability, Anuchin emphasized the need for nonexhaustive types of cutting and consumption of wood. With increasing frequency by the end of the Soviet period, specialists proposed that economizing was not simply about taking care of forests as a resource and raw material for the national economy and society but also about preserving a recreational asset serving society. Householders were urged to care about the future of the forests and hence themselves. Technical progress transformed the image of the value of forests: they were not only a source of wood to improve the quality of a growing consumer society but a factor for "improving the environment around the human" too.[70]

By the late 1970s, some called for the development of ecology-oriented industry; "one of new features of modern industry is its more ecological nature [*ekologizatsiya*]," as one person put it. Industrial technology, in this sense, was not only a means of industrial manufacturing but could provide more ecological types of wood harvesting and processing as well. New techniques and better methods were crucial here, showing how progress could positively influence industry-nature relations.[71] By the 1980s, the growing demand for wood was further constituted by specialists and high-level politicians as "an objective necessity. In the course of the economic, social, and cultural development of the people, individual needs for natural resources are growing up not only in terms of quantity but also in terms of diversity."[72] Addressing the global state of forests, specialist V. Kyucharyants stated in an article titled "Wood to Construction Sites," published in a regional newspaper in 1981, that forests were being turned into deserts through enormous levels of exploitation.[73] Forests remained "a colossal treasure of people, its national heritage [*dostoyanie*]," and conserving and enlarging forest riches was now "one of the main tasks of the Soviet economy."[74]

Late Soviet industrial legislation looks very environmentalist, even if quite formalized. In 1981, the Council of Ministers of the USSR decreed the imperatives to economize and make rational use of various resources, including waste, to save labor costs, materials, and capital investments as well as protect the environment.[75] Furthermore, "the Main Directions of Economic and Social Development of the USSR" for 1981–1985, looking forward to the year 1990, insisted on the need to employ the complex use of raw materials, resource-saving techniques, no-waste and energy-saving equipment, and various types of resources including used raw materials. In 1984, the Central Committee of the Soviet Communist Party and Council of Ministers issued a decree on the rational use of forests that held as its main aim the sustainable use of forests, citing the consumerist value of forests. Forests were not valuable until they were used by people. The logic underpinning this statement implied that if forest resources were not proposed for use, it would not be important to save them. From this perspective, sustainability was packaged into an industrial discourse and closely connected with production and consumerism. Yet forests were now seen not only as a source of consumer products but also a natural actor producing biodiversity.[76]

These decrees were not effective in practice and mainly remained "a law on paper." They were, however, important in terms of the conceptualization of a new philosophy of forest use, emphasizing that rationality in the exploitation of natural resources evolved initially from the need to make cheaper consumer goods, and gradually coming to incorporate views about the value of biodiversity and the recreational function of forests, while nonetheless remaining part of industrialist discourse. Prior to the end of the regime, new industrial legislation allowed the enterprises more freedom and made them responsible for the quality of production, as the 1988 decree on enterprises stated. At the same time, it appealed to the rational and complex use of natural resources, and invoked the mild negative impacts on the environment. In practice, though, it did not work. One of the most notorious examples was the Baikal'sk pulp and papermaking plant. Launched with unfinished water treatment equipment, the plant became a disastrous polluter of Lake Baikal. Another illustration was the wood harvesters of the Dal'lesprom industrial merger, which cut forests without permission for years.[77] But in the late Soviet years, some engineers stood for more ecological production, clearly connecting economic rationality and environmentalism. Interestingly, in the mid-1980s, professional forestry journals began to publish articles on environmental protection that highlighted measures that enterprises had taken for treating air and water pollution near production sites. In the 1980s, there were so-called technical committees that worked at enterprises to discuss the industrial impact on the environment at each particular production point, investigating local water basins, forest stocks, and so on.[78] In 1983, A. Kubenskiy, vice chief engineer with a specialization in wastewater, wrote, "The debt of every Soviet person is to take care and save nature" because nature is "our great treasure, the foundation of material production, the source of welfare and health."[79] He still saw nature through the lens of its material potential and first listed the industrial meanings of nature (to produce goods, for instance). Yet unlike earlier commentators, he stressed that Soviet people had to take care of nature as it was their civilian and moral debt.

Indebting the importance of forest protection to each person reflected a typical Soviet approach of making collective tasks a personal obligation. In line with Marxism, Soviet citizens were members of the collective, but simultaneously took personal responsibility for fulfilling (and overfulfilling) the plan and increasing the quality of manufactured products. From the perspective of this shared responsibility, every person, whether a specialist or not, was to take care of forests. But specialists themselves saw the role of experts as most crucial in developing a more environmental approach to nature. As an engineer said in 1990, "It was impossible to normalize the economic development without *purifying ecology*. It means that ecological aspects must be included in every scientific research and engineering project. Specialists—scientists and engineers—must have the weightiest say here. We must awake the civilian responsibility of those who work and create the technique already today."[80] These appealing words echoed the spirit of the day that followed "the storm over Baikal" in the 1960s, disaster in Chernobyl in 1986, and rise of Soviet ecological activism and environmentalism in general. By the end of the Soviet regime, some engineers used environmentalist rhetoric, underscoring that saving nature was an inherent part of industrial development and the main sign of progressiveness. This fixed a shift of focus from a perspective that understood the value offered by forests as purely economic to one that emphasized a higher degree of ecology in industry. As one commentator asserted, "The time when [a] forest was seen as a source of wood [only] is stepping away. In the first place is now the complex and multisided use of all the useful services of [a] forest and the making of sustainable forest use."[81]

In the 1980s, pollution and the contribution of deforestation to climate change became another ecological dimension added to the counting of losses. Many connected the irrationality in both the protection and exploitation of nature to negative environmental consequences. As noted by G. Vlasov, an employee of the Bratsk LPK, because of the enormous air and soil pollution of this enterprise, many forests "*died* and *were poisoned*" near Bratsk.[82] Simultaneously, the overcutting of wood to supply the industrial complex led to dwindling stocks of fir forest. Despite this loss, enormous amounts of waste were still left in forests and near waterways, contributing to a complicated picture of Soviet industry-nature relations.[83] Those supporting the use of waste also attributed environmental dangers to burned wood waste and waste thrown away, in contrast to the singularly productivist lens through which they had previously seen waste as an economically unprofitable practice.[84]

4

REED BECOMES WOOD VALUE

INDUSTRIALIZING "HOMELESS" PLANTS

It was not only waste products that specialists saw as offering alternatives for industrial wood supply. The industrial potential of straw and annual plants, they argued, had also not been fully utilized. Reeds or phragmites (*kamysh*, quite frequently also called *trostnik* in Soviet sources) are perennial grasses that spread in warm parts of the world and grow in river deltas. In the Soviet Union, they covered large tracts across Ukraine, the south of Russia, and Kazakhstan. Local dwellers traditionally used this plant for constructing houses and making forage for cattle. In the framework of the search for more rational resources for industrial production, annual plants showed great promise, with a chemical structure similar to wood, and could be used for cooking and manufacturing paper. Annual plants formed one component of the experimental model that specialists proposed as a solution to the wood crisis, and many suggested that they could provide a sustainable industrial supply of raw materials. In the context of late Soviet industrial development, new materials could, on the one hand, serve as a remedy for depleting industrial forests, and on the other, offer a more modern and economically efficient attributes. Reed was plentiful, seemed to be easily harvested, and was applicable for a wide range of industrially manufactured consumer products, such as low-quality paper and cardboard. It promised another avenue to acquire cheaper industrial material because its processing reduced time for industrial operations. Indeed, from an economic perspective, reed was cheaper than wood; in 1955, the input costs for making one ton of cardboard from reed were about 8 percent less than what was required to make the same amount of cardboard from wood.[1] And compared to wood, it was much more sustainable: while trees took at least fifty years before they reached a suitable size for industrial use, reed regrew every year.

The first attempts to use annual plants in manufacturing various consumer goods were made well before state socialism in the nineteenth century. In particular, some Russian manufactures used straw, old cloths, and fishery nets for making paper as early as in the nineteenth century, as was also practiced in other countries (especially Germany). In the early 1870s, industrialist M. A. Putikov established a paper manufacturing site in Astrakhan' in the south of Russia to recycle old ropes and fishery nets. The production remained quite limited, however—there were only nineteen workers and two steam-making machines—and the facility shut down in just a few years.[2] Such a small production was designed for manufacturing low-quality paper to supply unforested regions where industrial wood was in short supply. Initiatives similar to Putikov's had reduced production cycles because of the size of manufacture and did not operate for long. In the 1930s, some Soviet specialists spoke in favor of using similar materials in industrial production, but this project did not develop either, owing to weak technological infrastructures at Soviet cellulose and papermaking enterprises. The deficit of paper was obvious at that time, but the government tried to solve it by increasing the existing capacities of wood consumption along with annexing several cellulose, papermaking, and wood-processing enterprises after the Second World War from Finland, the Baltic states, and Japan.

In the 1950s, specialists suggested that experiments with alternative resources were a more progressive solution than the extensive enlargement of wood-based enterprises. Reed in particular was seen as a promising material in terms of industrial value and became the center of interest when many specialists spoke confidently about how modern industry had the appropriate technologies to process the material. Unlike earlier initiatives that had remained local in scope, these new enterprises were large-scale and state led. Postwar industry had a strong economic interest in industrializing annual plants as an alternative to industrial wood, most immediately to supply the southern regions of the country. Revived as a nation-scale idea, late Soviet enforced industrialization saw alternative resources as modern materials.

The prospect of having a rapidly replenishing raw material was not only a matter of technological experiments made by specialists but also attractive for high-level officials who were interested in achieving more voluminous production at lower costs. In the context of the Cold War and East-West competition in consumer production beginning in the 1950s, reed and other alternative raw materials particularly mattered: they were seen as the key to making an efficient system of the industrial consumption of natural resources, decreasing deforestation due to the enhancement of industrial production, and fostering a more sustainable use of forests. At that time, the search for more economical raw materials and alternative sources of energy became pivotal too.

Specialists spoke about the huge economic potential of reed as a source of numerous manufactured products. Engineer V. Mudrik, for example, calculated that one ton of reed could yield two hundred kilograms of feeding yeast, more than three hundred cubic meters of fiberboards, up to four hundred kilograms of cellulose, four hundred kilograms of paper, or up to six hundred kilograms of cardboard. This list measured the value of natural resources with a consumerist lens and presented modern materials in demand, especially cardboard, a crucial consumer material used for packaging food, shoes, and other consumer goods. As Mudrik proudly concluded after evaluating the stocks of this prospective plant, "This is what reed can really give us. It is thus a very valuable plant that simply grows in many our regions."[3] In his work, three ideas are particularly important in revealing the new meanings attributed to reed as a natural resource for industry. First, he mentioned an economic *value*, the term that defined a resource of nature as a participant in the economic process. Reed, not yet involved in the Soviet production, carried a *potential* benefit and yet remained excluded from the industrial cycle, thereby waiting to be involved in the action. Second, he noted that reed grew abundantly in many regions, mainly in the south of the Russian Republic, Kazakhstan, and Ukraine. As in the case of forests, reed availability was connected to notions of natural riches and economic abundance. Unlike wood, however, it was seen as growing more quickly and so was more economically profitable. As in case of forests, many stressed the imperial character of this natural resource when speaking about quantities: "The USSR has the largest squares of reed, which is the most valuable raw material that had significance for people's economy. It is the time now to reconsider our attitude toward the reedbeds."[4] Third, this view was similar to the professional outlook of waste: through the discovery of technical possibilities and recognition of the wood crisis, previously unutilized material gained a specific economic value and came to be tied to hopes for more rationalized industrial processes in the future. This sense of gained value exemplified a shifting model of industry-nature relations and involved "new" resources in the industrial chain to increase manufacturing where traditional wood was insufficient. At the same time, many stressed the enormous role of reed in industrial production: they saw it as an alternative remedy for easier harvesting and a way to foster a stable base for supplying numerous enterprises with raw materials. Saving forests from depletion therefore emerged alongside productivity aims as a by-product of the industrialist vision of nature.

Reed was just one example that emerged in the multiple searches for alternative materials. In the mid-1950s, many specialists sought technologies to efficiently transform alternative resources into new industrial raw materials. The publication of books and brochures that claimed to have "discovered" new types of materials boomed. In general at that time, engineers and scientists working at different institutions discovered new characteristics of raw materials. For instance, a few engineers at the Leningrad technological institute for the pulp and papermaking industry reported on experiments with new types of pulp at the Kherson pulp plant in Ukraine, explaining that new sources, including aspen wood, were progressive.[5] Aspen was not widely used in industry, but it offered appropriate qualities for increasing everyday consumer modernity; toilet and other sanitary papers made from that wood species were milder and cheaper, and so more readily available to consumers.[6] Specialists used the word *recycling* (*perepabotka*) when speaking about both recycling wastepaper and processing the annual plants used in industry. They saw plants as having definitively changed their meaning from waste—in terms of their potential value and something previously not demanded by the industry—to an industrial material. Technology was to industrialize plants never before used in production at a large scale. Reed was thus elevated from domestic to industrial use.

EXPERIMENTING WITH NATURAL MATERIALS

In June 1955, the minister of the paper and wood-processing industry, Feodor Varaksin, and his vice deputy, Nikolay Chistyakov (a multiply awarded Soviet official and forestry specialist), decreed that the industry would conduct scientific research on reed, which grew abundantly in the delta of the Volga. There were some immediate investigations of the region conducted by local scientists and engineers in Ukraine on the decree from the political center. In particular, in 1955 the Ukrainian branch of Giprobum, the key planning institution of the paper industry, undertook research on reed in the region. Its main revelation was that phragmites formed one of the most widespread plants "stretching from the tropics to the polar circle" and were an industrially appropriate raw material. This was a positive conclusion that once more replicated the image of the Soviet Union as a country abundant in natural resources; "reed," the Ukrainian branch wrote, "grows almost in every republic of the Soviet Union and covers huge squares [swathes]. In the Asian republics [of the Soviet Union] alone, mainly in the Kazakh SSR, the square of clumps is about two million hectares."[7] In other words, the Soviet Union was not only full of wood, oil, gas, coal, and other well-known natural resources but also reed, a long-neglected source of industrial value.

In Russia, reed grew in the deltas of the Volga, Kuban', and Don, in the lower course of the west Siberian plain on the bank of the Irtysh, Ob', and Yenisey, in the Novosibirsk and Omsk regions, in the Far East, and in Ukraine in the lower course of the rivers Dnieper, Dniester, Danube, and others. Specialists often reported on these geographies as holding national riches underfoot that could be extracted more easily than cutting wood, and could provide quicker manufacturing and cheaper products of the equal quality. Hence resources that offered alternatives to wood were given priority as the levels of their consumption remained miserably low. Using other natural resources instead of cut wood was to become a matter of urgency. As the ministerial report on the investigation of Volga natural resources in 1955 claimed, the "annual harvest of reed in the Soviet Union makes up, according to the most *humble* calculations, more than thirty billion tons, from which the economy currently consumes just a miserable portion." The report also referred to Khrushchev's speech in which reed was heralded as "a wonderful material."[8]

The 1955 report explained that local people in the regions covered with reed had been the main consumers of this plant, which they had used for construction, fuel, and heating their homes, utilized as a natural alternative to wood, coal, and peat, which were all lacking in supply. The document demonstrated that the ministry was particularly concerned with the sustainable growth of reed to facilitate continuous industrial production. As the report said, "The question about the regrowing of reed is important, especially because we have knowledge about the Korean and Chinese experience of destroying reed after harvesting" (the report provided no detail about that experience, however). It warned, for instance, that the construction of new hydropower stations in the delta of the Volga could damage reedbeds. What it proposed, in fact, was translating local values of reed into industrial scales, converting limited uses into voluminous manufacturing. The ministry sent a request to industrial institutions in the south of Russia asking them to examine the existing practices of harvesting and growing reed. One replied that according to their observations and talks with local users, the "annual harvesting of reed not only does not hinder but instead favors the sustainable growth of reed."[9] The report concluded that the sustainable growth of reed could provide a resource base for industrial development.

In 1955, the Institute of the Paper- and Cellulose-Making Industry requested that the ministry send its specialists into a few countries that had some experience using alternative resources instead of wood. It requested a visit to East Germany to study the technology of cooking straw used in the city of Wittenberg, inquiring about sending "the whole brigade" there as the USSR had no experience of developing such technologies. The institute also proposed sending specialists to China—to examine how the Chinese cooked pulp from annual plants, such as reed and rice straw—and Italy, the Netherlands, and Sweden to study *if* there were similar experiences and "collect the maximum of materials for choosing the most advanced technology for processing annual plants."[10] Documents show that Soviet institutions did not have a clear vision of foreign enterprises that used alternative nonwood materials but rather *implied* that industrially advanced economies potentially could develop these projects. Archives do not reveal if these travels really took place, but Soviet specialists saw foreign industries as effectively using natural resources, which for them meant extracting the maximum benefit that these resources could give. Importantly, specialists had theoretical rather than practical knowledge about using reed and sought out any experience abroad. They did not mention the prerevolutionary experience, despite some czarist scientists having previously examined and described reed. As early as 1840, for instance, G. Kuzmishchev described the huge reedbeds of the Volga and their economic application in heating brickmaking factories in Astrakhan'.[11] Yet Soviet specialists were particularly interested in the large-scale industrial use of reed, evidencing a break with the pre-Soviet past.

In 1956, a group of Soviet specialists from several institutions of the Russian Republic and Ukraine traveled to Romania. In the report they wrote on their return, they noted that they had established contact with all the administrative employees and institutions that worked on the use of reed in industry. Soviet specialists visited reedbeds in the delta of the Danube, a cardboard-making factory that was under construction to process reed, and a factory that produced harvest and transport machinery. What the Soviet delegates found in Romania was similar to what had existed in the Soviet Union at that time: there was no ready reliable reed-harvesting machinery in use; Soviet visitors, they evaluated, had better expertise and had more thoroughly investigated the qualities of reed as compared to their Romanian specialists. As they wrote in their report, "Our group, to the extent our knowledge allowed, tried to help Romanian comrades to see weak points in projecting and constructive use of machinery and mechanisms."[12] Comparing Romanian and Soviet research, they also admitted that Romanian construction was slow, the technology for using reed they chose was not "progressive," and investments became a heavy burden as they significantly increased because of the low productivity of the project. Contrary to this experience, many volumes published in the 1950s and 1960s about reed and its industrial potential enthusiastically talked about the Romanian experience and perspectives.[13] Interestingly, Soviet institutions tried to include the industrial use of reed and other annual plants as a topic for investigation within joint research initiatives of the socialist bloc.

Despite strong enthusiasm about the potential of reed, ministerial producers admitted that harvesting the material was complicated primarily by technical factors: reed grew in water, which required new equipment. In the 1950s, the industry in fact harvested less than 20 percent of all growing reed because of, as reports explained, a lack of machinery. Some wrote that "it evokes hesitation about the sustainability of raw material base [from reed]." They also questioned

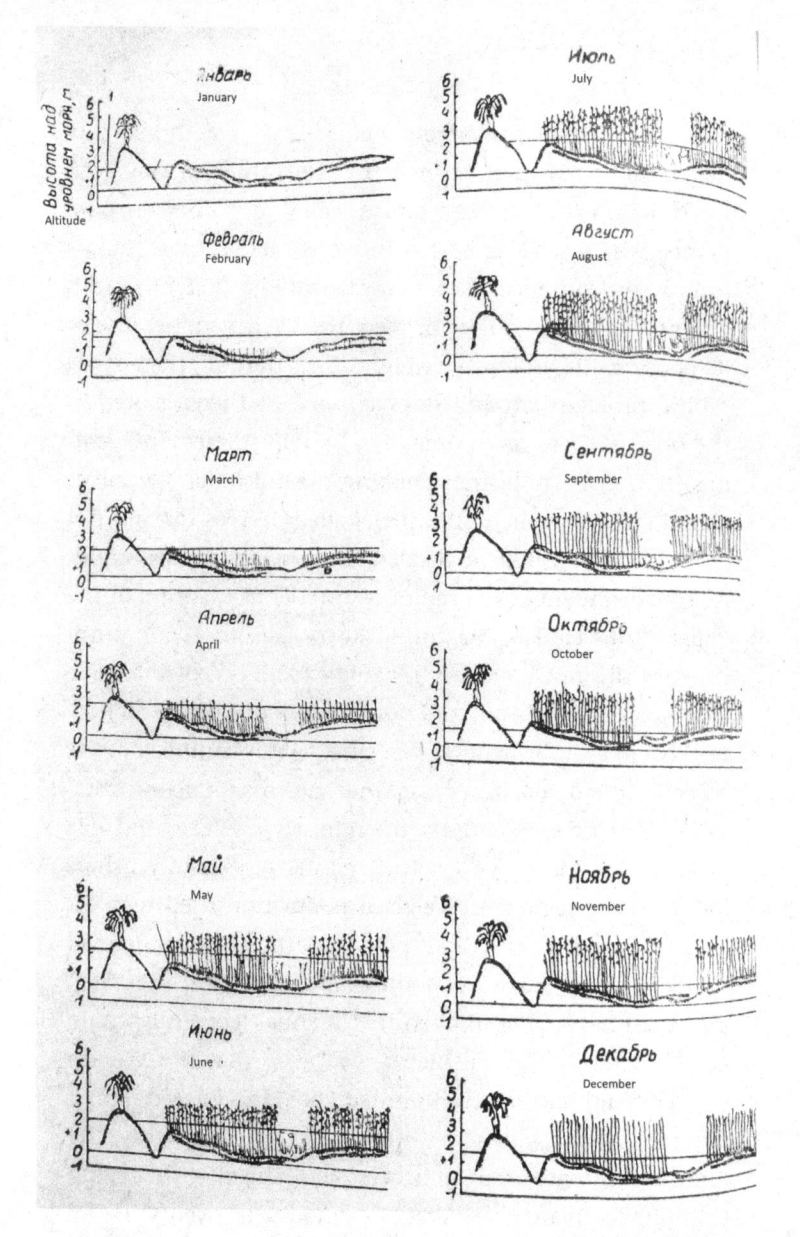

Figure 4.1 Soviet scheme depicting the growth of reed from January to December. *Source:* F. F. Derbentsev and A. F. Grishankov, *Zagotovka i khranenie trostnika v Rumynskoi narodnoi respublike* (Moscow: B.i., 1959).

the reed's capability to regrow every year if it was industrially harvested. The expert warned that the situation was even worse in the areas of the future Ismail and Kherson pulp plants, which were projected for construction to manufacture paper from reed. Soviet-constructed harvester machinery was instead a danger for reed, they believed, because of its poor quality and heavy weight.[14] In addition, there arose typical problems around storage: harvested reed, stored in the open air, was often drenched by rain, thereby preventing its use for industrial cooking.[15] Ministerial specialists complained that the authorities issued decrees but did not investigate anything in detail before making a decision.[16] As Feodor Kuteinikov, a senior researcher at Institute of the Paper- and Cellulose-Making Industry, said in 1960, "Until now we did not have any reliable experience of intensive exploitation of reed and did not have scientific data on how reed could grow again after planting, so this complicates the construction of machinery and mechanisms for planting it." When looking for solutions, the ministry proposed the idea of attracting local dwellers to harvest reed—mainly fishers living in the region. If the local population used reed for its own purposes in decades past, now the ministry defined its reedbeds as "zones of industrial exploitation." At first, industrial harvesting implied that manual labor still would be practiced. It was calculated, however, that the average worker could harvest five hundred kilograms of reed manually during an eight-hour working day. Consequently, the ministry estimated that to harvest all the reed of the region to fulfill the plan, the incredible amount of twenty-five to thirty thousand seasonal workers would be required.[17] In 1962, the state-led harvesting of reed constituted twice the amount that locals achieved altogether during the year. To organize the large-scale harvesting, the ministry employed more than seventeen hundred permanent workers and more than five thousand seasonal ones. Thus while large numbers of people were involved, this was still less than was required to substitute mechanical operations with manual labor.[18]

Until that point, the ministerial harvesters had never had any experience of growing reed, while local workers did not have any experience of the industrial growing of reed. The main differences between the industrial and domestic use of reed lay in scale as well as types of storage for the harvest. The domestic harvest of reed was on a much smaller scale and undertaken with more care, as dwellers harvested it manually and did not use heavy machinery, which could destroy the shoots. Soviet-made tractors used for agriculture were heavy and not appropriate for working in water. The purposes of industrial harvesters were ambitious, though: the ministry proposed to grow what it considered more 'progressive' types of reed (e.g., Mediterranean reed), which it believed had a higher productivity and could produce more voluminous harvests. Yet this proposal was left unrealized.

The project of using annual plants revealed a tension between, on the one hand, the imagined picture of the progressive use of reed as an alternative to wood supply for unforested regions, and on the other, practical implementation, which suffered from a lack of skills, knowledge, and technological infrastructure to use the material industrially. The initiative to industrialize annual plants was a reflection of modern industrial processes when greener production mattered as a rationalized practice. Despite obvious problems with reed harvesting and the lack of relevant knowledge about it, the Soviet leadership launched the construction of a network of enterprises to process this new and, as many still believed, promising material.

INDUSTRIAL BUILDING IN THE SOUTH

In 1963, G. Asteryakov, the head of the factory committee of the newly built pulp and paper plant in Astrakhan' in the south of Russia, wrote that the enterprise was "new in all senses. . . . A big life came to the shores of the delta of the Volga [River]."[19] Historical sources attribute the decision to create a network of industrial enterprises to process annual plants to political will at the highest level, which supported professional proposals for the use of reed. In the early 1960s, one political memoir explains, Khrushchev and the former minister of the forestry industry and vice president of Gosplan, Georgiy Orlov, were flying over the delta of the Volga. Orlov asked Khrushchev to look out through the airplane window at the vast areas covered by reed, noting that it might be a good place to construct a large pulp and paper plant. Inspired, Khrushchev immediately gave an order to Alexey Kosygin, the head of Gosplan, to build such a plant near Astrakhan'. Kosygin expressed his hesitation, stating that Gosplan had already approved the list of new plants to construct. He also said that he never heard about manufacturing pulp and paper from reed. But Khrushchev insisted on immediate construction. After a short squabble, Kosygin conceded and the plant was built.[20] The Soviet leadership came to include reed as a raw material in the five-year plan. As Khrushchev said in one of his official speeches in 1963, "Reed is *profitable*: today you cut it and then the next year it will grow again."[21] He stressed that the industrial potential of reed had been underestimated and this was the reason for primitive methods of harvesting. Through the political lens, reed was thus an industrial miracle to supplement wood that was costly for the south. In the course of rapid industrialization in the 1960s, three enterprises were constructed in the Russian, Ukrainian, and Kazakhstan republics. Among the projected enterprises were several more in the southern part of the country: the Astrakhan', Kzyl-Ordynsk, Ismail, and Kherson plants.

Soviet planners aimed to build a network of reed-based enterprises to launch the cheap and sustainable production of pulp, cardboard, and low-quality paper from annual plants in unforested regions in the south. The harbinger was the aforementioned cardboard-making factory (later enlarged into the pulp and cardboard-making plant) in Astrakhan'; its construction was launched in the early 1960s. It was completely equipped with Finnish machinery, such as special cookers for reed.[22] The construction of the Ismail pulp factory near Danube in Ukraine began in 1963. In addition, the Kherson pulp plant was built in Tsyurupinsk in Ukraine by the mid-1960s. The construction of the Kzyl-Ordynsk pulp and cardboard-making plant also started in Kazakhstan in 1964. In addition, a few more similar enterprises were projected for the coming decade. All of these projects were part of industrial exploration of the southern regions of the country, echoing the unspoiled lands campaigns where exploring was held as more crucial than long-term planting on new lands. The plan was grandiose: to launch at least four enterprises fully supplied with reed to satisfy the needs for paper and other products in the unforested south. Specialists believed that the cost of the construction would be returned in just one year due to the fact that products made from reed would be much cheaper than those made from wood and the production itself would be more sustainable.

Yet the question of reed availability in the future was among the most pivotal since the initial days of the construction. Many reports drafted by ministerial and research experts concluded that the real availability of reed at industrial scales was a bigger problem than the promising perspective anticipated. Paradoxically, these pessimistic prognoses were advanced simultaneous to the construction and the enthusiastic praise espoused for the potential of reed by political leaders. Thus the 1962 expert conclusion on the general harvesting scheme in the Astrakhan' region insisted that the amounts of reed were not sufficient to satisfy all industrial needs.[23] Specialist Zaitsev wrote that the change of the raw material base had been caused by intensive building on the banks of the Volga, which became a fully regulated river controlled by a cascade of hydropower stations, leading to a change in the hydrologic regime of the river. This decreased the quantities of reed available for making cardboard and meant its further degrading. As Zaitsev concluded, artificial interference in the river would lead to the loss of the industrial potential of reed and the enterprises built nearby. The industrial potential of reed was most crucial for him, and he measured the environmental destruction from the perspective of the capacity for industrial production. For example, the Kakhovsk and Kremenchugsk water reserves, opened as exemplars of Soviet industrialization, flooded the reedbeds: "The experience of harvesting reed with machinery in flooded areas showed that the damage increased by

50–60 percent."[24] The main problem lay in the fact that the constructors could not build lighter machines and the weight of the mechanisms they had available destroyed reedbeds. The paradox was in the fact that lighter machines were not usable in the delta of the Danube because their low-capacity motors and other weaknesses led to frequent breakdowns. The percentage of decimated reed roots in some cases exceeded 80 percent after chain tracks flattened and treaded the plants. The reedbeds ruined by the tractors could only be revived after several years.[25] Because the harvesting proved so damaging to nature, specialists suggested ridging and irrigating—forms of technical construction to prevent the flooding of reedbeds and increase productivity. They hoped that it would increase the harvest, doubling the yield of each hectare, but the project itself was costly.[26] Attempts to launch cheaper production chains of traditional wood-based products from alternative raw materials thus devolved into an expensive quagmire since the harvesting and cooking of reed required expensive machinery, infrastructures, and technologies.

In 1961, the ministry built special agricultural and biological stations in Tsyurupinsk on the Dnieper and Ismail on the Danube to investigate reed, particularly its harvesting and regrowth. In the early 1960s, several reed-harvesting stations (*kamyshezagotovitel'nye stantsii*) were also established on the Volga. In 1964, the authorities established special reed-harvesting stations in the deltas of the Dnieper and Danube called *kamyshpromkhozy*, similar to traditional lespromkhozy.[27] As early as 1964, however, several reed-harvesting stations were closed in the eastern part of the delta of the Volga because they stopped producing output;

Reed harvester ZhBT-1.8

Reed loader with a car

Figure 4.2 Soviet experimental machinery for reed harvesting. *Source:* Lazar' Kanevskiy, *Zapasy trostnika v SSSR* (Moscow: B.i., 1965).

reedbeds there had been all but destroyed by mechanical harvesting. The authorities sadly became prisoners of their own decisions: exploring new areas of reed cultivation and establishing new harvest bases, they had to construct housing and social infrastructure for workers, which proved a costly enterprise. The Volga hydropower station had a negative impact on reed too: it periodically flooded the plants. Harvesting nonetheless remained the primary challenge, as the "losses from production are not big, but the losses from the reed-harvesting practice are enormous."[28]

Immediately after the first stage of construction was completed, the lack of reed became a real danger—not only for nature, but in terms of fulfilling the industrial plans that as in the whole Soviet economy, were seen as a matter of paramount importance. The cost of harvested reed and manufactured cardboard was three times higher than had been projected, while the quality of production did not correspond to the requirements. The enterprises received less than half of the reed required and satisfied less than 50 percent of the need for raw materials.[29] Initial expectations to develop an alternative base for industrial manufacturing were in this way dashed, revealing that the reed project could not offer an alternative path for the development of products traditionally made from wood in the existing Soviet infrastructure.

RICH AND POOR NATURE

As early as the mid-1960s, simultaneous to the construction of reed-based enterprises, some specialists and politicians of the highest level expressed serious concern about the lack of raw materials for new production. They were concerned about the destruction of reedbeds not because it entailed the annihilation of ecosystems but rather because it held industrial consequences. In 1964, some claimed that despite it being the midseason of harvesting, only 22 percent of the harvest had been completed for the Astrakhan' plant and slightly less than 30 percent of reed had been harvested for the Kherson pulp factory located in Ukraine.[30] The situation in Kherson was most difficult: the harvest of reed there was almost zero, and the managers of the factory had to purchase raw *wood* from forested regions that was transported over a distance of a thousand kilometers. By the mid-1960s, only one-third of raw materials was constituted by reed, while the rest was supplied by wood from the north of the country. This led to disappointment among many specialists about the prospect of building new reed-consuming enterprises, as had initially been planned.[31] Moscow, however, was more enthusiastic about the current and future situation around the construction, and insisted that planned enterprises could not be canceled. The reason for that lay in positive expectations: moved by industrialist expectations, the planners believed that reed could be artificially flooded and its regrowth productivity thereby increased.[32] They expected flooding to make the biotopes more homogeneous and allow for reed to grow faster. Some scientists supported the idea of flooding reed, particularly those from the Institute of Botany at the Kazakhstan Academy of Sciences who worked out a plan for flooding reed. These initiatives happened despite protest from other foresters and harvesters, who said that the planned flooding of half of all reedbeds did not correspond to the water regime and historical conditions of the riverbeds.[33]

Simultaneously, the factories already built were constrained by the plans and ended up asking for supplies of wood instead of reed—an unstable material, as they put it, which was experienced primarily as a problem rather than a promise.[34] In 1966, workers of the Kzyl-Ordynsk explained that the enterprises did not work well because of the lack of raw materials: the factory produced just under 80 percent of planned cardboard, but only about 24 percent of packages.[35] This is why some planners spoke about the need to purchase Western equipment to produce materials from wood as opposed to reed—ironically at the enterprises precisely geared toward reed processing. The situation was complicated by the rapidity of construction; three factories were built without serious estimates of risks. The planning system managed to count the quantities of production, but did not assess the risks and pitfalls in the haste to compete with the West. This story also reveals that Soviet planners had an obscure vision of how closely raw materials, environment, technology, and industrial production were connected. And archival documents show that industrial and research institutions did not know how many natural resources were in fact available.

In the mid-1960s, the local authorities of reed-consuming enterprises planned to combine supplies of raw materials, mixing both reed and wood as a form of compromise in the face of reed shortages. From their perspective, supplying the enterprises with raw resources and making appropriate machines would take between five and seven years, prompting the necessity to supply the reed-based enterprises with wood and avert delays in production.[36] In 1964, the initial project of supplying the Astrakhan' plant was changed in order to supply it with wood like other forestry-based enterprises. Interestingly, plans for new enterprises on the same lines were not scrapped, and construction continued even when it became clear that future supplies of reed would not be possible. Supplying wood for factories in unforested areas was costly, requiring lengthy transportation from the northwest of Russia and Siberia, and involving the hiring of a range of specialists. The planners noted the advice of scientists and engineers—experts who explored the availability of reed—but they regarded negative prognoses as temporary, and insisted that the situation would be improved through rational mechanisms for growing reed and appropriate harvesting machines. The ministerial lobby for the industrialization of the south was strong and held fast to the belief that the expansion of the resource base would be possible after the completion of the factory construction. Meanwhile, as they believed, partially supplying the plants with wood would help launch production without having to wait for technological achievements to increase the reed harvests. The Soviet system admitted no mistakes or uncertainty, and this is why the planners did not predict the multiple problems on the horizon with accuracy.[37] The change that the planners undertook in 1964 concerned the profile of the enterprise in Astrakhan', shifting from cardboard to print papermaking as the demand for the latter was larger. This shift was necessitated by the change of raw material supplies. In 1964, the Moscow Giprobum, the organization responsible for the construction in Astrakhan', suggested that the enterprise would make only 65 percent of the products from reed and 35 percent from wood and sawmilling wastes. This combination enabled the factory to make consumer paper.[38]

For the Ismail plant in Ukraine, in accordance with its plan, before the availability of advanced harvesters, reed was to be harvested manually only in order to supply new production. This in fact contradicted Soviet plans of the 1960s for launching fully automated enterprises, in line with the proclamation that mechanization and automation should become the key elements of the socialist project.[39] In fact, what was witnessed was the reverse process: humans came to substitute for the machine at modern enterprises, proving what Jenny Smith called "involution" in Soviet agriculture.[40] The Kzyl-Ordynsk pulp and cardboard-making plant faced the same problems. As the chief engineer of the Kzyl-Ordynsk reed-harvesting enterprise, V. Karymsakov, wrote, "Because of the worse situation with flooding reeds, their squares and productivity of bed rushes significantly decreased, and it creates additional problems for supplying the plant with raw materials."[41] The lack of reed obstructed Soviet modernization along with the hopes of moving to material prosperity and satisfying consumer production.

The problem of supplying enterprises with raw materials was so pivotal that scientists at various research institutions undertook intensive experiments to improve the productivity of reed. In Astrakhan', the main institution that examined reed was the Central Research Institute of Reed, established in the early 1960s. It proposed growing reed together with fish and to "assign" workers of industrial fisheries with the responsibility of reed growing. In other words, scientists looked for options to make harvesting cheaper and sourcing labor resources that were in short supply in the reed-harvesting sector.

The amounts of usable reed remained limited, though, and experimentation did not open any avenues for the increase of the harvests. As such, reed-consuming enterprises began dividing available reed among themselves, while each received additional supplies of wood—even shortly after launching. Further compounding the problem was the fact that the equipment installed there could not be used for processing wood, so using additional supplies of wood to keep the production sustainable was not possible. As a result, in 1967 some shops stopped working at the Astrakhan' factory. One that produced fiberboards ceased production after it did not receive reed and could not manufacture fiberboards that met Soviet quality standards. Counting the real supply of reed, engineer V. Bogdanov inquired about limiting the amount of fiberboards required of the shop to six million square meters, but planners more than doubled that figure, demanding fourteen million square meters per year. The decision had been made at the highest level based only on economic interests and the abstract logic of intensive production rather than on grounded concern over the destiny of reed. The highest-level institutions—the Central Committee of the Party and the Council of Ministers of the USSR—made corrections for the production plan of the shop twice, increasing the quotas significantly.[42] Construction was undertaken in accordance with that plan, but the real raw material base did not match the plan and was unable to facilitate the demanded production.

The problems associated with creating appropriate infrastructures lasted until the end of the regime. Thus usable harvesting machines were never actually built, despite many attempts. After the 1970s, the question of reed-based production fell into neglect, and wood almost completely substituted reed. Yet specialists working in forestry became more critical toward the industrial methods that led to the destruction of reed, mainly because of disrupted industrial plans. The reed-based enterprises faced interruptions in cellulose making because of the lack of resources, along with the negative industrial practices typical of the Soviet system, such as the rapid turnover of labor and a lack of repair workers. Because the enterprises were completely equipped with Western machinery, as in many other enterprises across the country, "the repair parts were not produced by Soviet industry [even as] the limits on purchasing foreign equipment were small." This is why "this expensive equipment often broke down and did not work for a long time," resulting in dropped output.[43]

Annual reports on the activities of reed-based enterprises show that in conditions of raw material shortages, they had to search for alternative sources and methods to fulfill the plan. By the late 1970s, the Astrakhan' enterprise was being supplied with 60 percent wood and 40 percent reed. Reed harvests dwindled even further; in 1979, the harvesters supplied half the amount of reed that was required, even though the planned amount was small. In February 1979, the situation at the Astrakhan' plant grew especially critical because of the shortage of raw material. To continue production in the main shops at the very least, the staff of the factory's wood storage unit went to the Volga to break the ice and lift logs that were lost while floated. As the director of the enterprise wrote, "Hard organizational and preparation work was conducted in winter to prepare the fleet for lifting lost logs," which they previously saw as rubbish. They continued lifting sunken logs during the whole year and beyond.[44] This practice had been used as an urgent measure since

Stalinist industrialization in the 1930s, when enterprises lifted sunken logs to use in industrial operations to make up for shortages.[45] As worker A. P. Borisova said in January 1980, the "plant has a fever because there is no raw materials."[46] Fever was a word frequently used in archival documents and periodicals that spoke about a specific style of work in Soviet enterprises: long stoppages due to material shortages and frenzied searches for resources according to rapid tempos often took place at the end of the year to fulfill the plan. The history of the reed-based enterprises in the last decades shows a constant struggle for raw resources to supply the production suppressed by the plan. The enterprise reports often described the situation as "very difficult."[47]

The reed project exemplified the Soviet tendency to render natural resource wealth into degraded assets while still failing to fulfill the industrial plans. As the case of reed use demonstrates, despite possessing rich natural resources, the Soviet planned economy failed to build a viable industry that fostered the sustainable use of raw materials. While in practice the project resulted in the destruction of reedbeds, the very idea of using reed had been motivated—as with the consideration of other alternative resources in industrial production—by an attempt to achieve the sustainable use of natural resources and save wood from depletion. Specialists' expectations around the regrowth potential of reed were high, and in their view, sophisticated technologies were to play a major role in exploiting newly discovered resources. The modernist vision of reed as a quick and cheap natural treasure that could facilitate modern consumption and sustainable economic growth was intended to save costs on transporting wood. Less wood and more annual plants that could grow more quickly would be consumed. This was a strong and well-articulated discourse that nevertheless remained largely a feature of the dreamscape of specialists, once again exposing the gap between professional expectations and infrastructural disabilities. This gap intensified ecological criticism directed against Soviet industrialism, contributing to forms of industrially embedded ecology. By this time, specialists and journalists clearly connected the shortage of reed and the function of southern pulp, papermaking, and cardboard-making enterprises with environmental risks. In 1970, the newspaper *Forestry Industry* published an article titled "Reed Sings, or the Delta of the Volga Is in Danger," warning that the Astrakhan' plant had been built in a hurry and without proper water treatment facilities.[48] The southern projects thus constituted a failed rationality, contributing to critical visions of Soviet industrialization and the increased value of ecology.

5

THE PRESTIGE OF MODERN TECHNOLOGIES

HOPE IN MODERN WOOD PROCESSING RISES AND FALLS

In the technological dreamscapes of late Soviet specialists, modernization as the means for increasing the industrial productivity played a prominent role. By modernizing industry, they meant, on the one hand, the upgrading of machinery and equipment, and on the other, the more efficient use of raw materials and natural resources. Modernization was frequently used as a synonym for *technical improvement*. Specialists strongly believed in the power of modernization as a long-run technological process that required stimulation and investment, and widely employed the terms *modernization* and *modernize* when discussing machinery, facilities, and technological operations.[1] In particular, in the forestry industry, nature and economy were to be connected through sophisticated technology. Improved technologies were to be put in the service of the mastery of nature, helping to make its exploitation less wasteful through more productive wood-harvesting and wood-processing techniques.

Equipping the industry with new technology and developing new methods of production (such as assembly lines and the continuous cooking of cellulose, among many others) had been discussed from the first decades of Soviet power and derived from a drive for rapid industrialization. After the Second World War, technological modernization received a new impulse from chemistry, became both a powerful industrial branch in itself and source of useful knowledge to change the approach to wood processing in industry. From the mid-1950s on, chemistry grew to become a crucial component in Soviet plans for building an industrialized and rationalized society, and contributed to the making of modern foods, clothes, furniture, construction materials, and other consumer products. Many specialists believed that chemistry could change patterns of consumption, bringing society up to international standards of material consumerism. Wood played a particularly important role in this transition: it was a material that through chemical additives, promised widened applicability to transform the material world that surrounded people.

By the mid-1950s, chemistry had become part and parcel of professionals' visions of the future development of forestry. Engineers emphasized the use of chemical methods for improving the processing of wood, once again connected to the discourse of rationality in resource use.[2] Chemistry was one of the key clusters of innovation in developed countries in the postwar period, and became one of the main spheres of progress given its potential to change the range of products along with the ways people, industries, and agriculture consumed.[3] Soviet ideologists placed chemicalization into traditional socialist incentives in the struggle for progress and the goal of building a modern society characterized by consumer choice. This type of society would no longer be dependent on harvests or incapacitated by supply shortages, but instead would consume modern products such as paper and plastic packages and synthetic cloths produced from cheap, highly processed materials that would be easily available. Politicians often explained these improvements as laying the material basis of Communism, a stage of development they sought to reach in the nearest future.

Beyond ideology, the material basis implied an effective way for developing industrial enterprises to improve material life and increase the consumption levels of people. It resonated with the growing emphasis on consumer society as a modern way of life—a model that took root in the United States, and a little bit later, Western Europe. Behind the Soviet rhetoric about reaching Communism, industry positioned consumption as the central driver of the goal to improve living standards, which compared to the prewar and immediate postwar periods, had already expanded significantly. Chemicalization was to increase the productivity of wood as an industrial raw material with a wide applicability in consumer industry and formed part of the modernization solution for the wood crisis. Specialists explained that chemistry provided the tools for achieving the rational and cheaper use of "natural riches" and various industrial wastes because synthetic products, such as silk, were much less expensive than natural ones. A revolutionary method introduced to forestry chemistry, for instance, entailed the manufacture of wood plastic from wood and sawmill waste—a material that could be used in both industry and homes. This signaled part of the "synthetic polymers revolution" in which wood plastic came to be used in consumer industry more widely to manufacture various products. Adding chemicals into wood processing also helped transform wood into bright white paper and firm viscose cloths, among other products. As the 1970 book *What Chemistry Makes from Wood* explained, the combination of chemicals and wood promised to produce a vast range of goods to furnish the whole apartment and create the material infrastructures to facilitate everyday practices. The book especially provided an interesting example of children's toys made from wood-based semisynthetic plastic, captioning a published picture as "the first meeting with celluloid," describing how a child first encountered a plastic toy manufactured by the chemical forestry industry.[4] In this sense, the toy made from wood plastic was a product of modernity and a desired material object, symbolizing the interaction between humans and technological progress. The chemical industry itself became a large consumer of wood, while chemicalization became an inherent instrument in technological progress and people's comfort.

The consumption of industrially manufactured modern products implied increased demands for raw materials, such as water, meat, stone, sand, and wood. These were conceived as materials of modernity extracted from nature and processed by science and technology to satisfy the standards of modern life. Importantly, wood plastics and other goods could be made not from wood as such but rather alternative resources such as wood waste, thereby demonstrating that chemistry could offer solutions for the wasteful forestry industry. Specialists promoting the modernization model believed that by minimizing waste and increasing the volumes of produced goods, chemical forestry could help solve the problem of wood scarcity. At the same time, the proponents of this solution did not deny the need for the

THE PRESTIGE OF MODERN TECHNOLOGIES 135

Figure 5.1 "The first meeting with celluloid." *Source:* Anatoliy Averbukh and Kseniya Bogushevskaya, *Chto delaet khimiya iz drevesiny* (Moscow: Lesnaya promyshlennost', 1970).

colonization and experimental options to solve the wood crisis—the approaches explained in the previous chapters. Many of those who insisted on technological modernization, for example, explained that a chemicalized, mechanized, and automated industry would waste less resources, instead processing more waste and low-quality materials. They thus supported the idea of a no-waste economy, while those who saw the eastern lands as the drivers of economic advancement also hoped to do it efficiently with the use of modern technology. In their professional dreamscapes, chemistry was to serve nature positively, preventing its devastation while allowing for its effective exploitation at the same time.[5] Overall, modernization was to result in more efficient manufacturing along with the rationalized use of raw materials, and in this way help shield nature from total depletion.

In 1953 and 1960, respectively, the Soviet government issued two decrees that stated the need to close the gap that saw the USSR "lagging behind in the forestry industry," and emphasized the importance of the "struggle for making the industry an advanced branch of the national economy."[6] Following these decrees, the first factory for manufacturing modern papermaking machines was constructed in the USSR in 1964. Such machines had previously only been purchased from abroad. At the same time, while stressing the crucial need to produce world-class paper products, specialists typically referred to Western samples as the standard to reach. Many wrote that in other countries, capitalist and socialist, furniture making and other branches of the forestry industry used numerous chemical products, such as synthetic rubber—materials still in short supply in Soviet industry. For this reason, specialists evaluated the industry as backward, comparing the development of socialist forestry with that of developed economies. They typically attributed this lagging behind to low technical levels along with the absence of cheap, productive, and reliable machinery.

Comparisons between the "advanced West" and "backward Soviets" featured frequently in Soviet industrial language, and in the case of forestry, the problem of more efficient manufacturing met that of resource exploitation and the imperative to save on raw materials. Raw material shortages always presented significant costs to the planned economy due to irregular supplies caused by the poor logistics typical of state socialism. Specialists therefore believed that chemistry could enable the industry to efficiently manufacture goods to provide the kind of material modernity to state socialism that was evident in capitalist consumer societies.

While chemistry offered numerous possibilities for making consumer products, and specialists underscored the need to invest in forestry chemistry, the manufacturing of some key goods was a source of constant problems. The production of paper and paperlike materials—one of the basic products of forestry chemistry—remained a pivotal issue for Soviet industry during its whole history. Since the Bolsheviks had seized power in Russia, they had tried to solve the problem of shortages of paper, the material required for numerous social and cultural projects such as the literacy campaigns launched in the 1920s. While Soviet leaders proclaimed that the Soviet Union was "the top reading country in the world," and promoted the publication of a large number of books and journals, the industry suffered from a lack of consumer paper. The gap between the political campaigns and technological possibilities was captured by an employee of the Institute of the Paper- and Cellulose-Making Industry: "In order to publish a brochure, we will find the author but will not find paper."[7] Admitting this fact, they stressed the gap between the intellectual capabilities of socialist society and its material infrastructures, and concluded that the way that scarce paper was being used was irrational. In 1955, the average consumption of paper per person was ten times lower than in developed capitalist economies: according to some calculations, while the average in the United States was 162.3 kilograms per person annually, in the USSR it was a meager 12 kilograms per person. Interestingly, some Soviet commentators attributed this gap to the fact that people in the United States needed more paper for printing advertisements, while "the socialist system of economy and Soviet trade do not require advertisements that the Americans do."[8] Behind this, however, the gap in paper and cardboard consumption was held as a testament to the differing living standards that marked the two societies, and the shortage of paper proved a serious problem in the Soviet economy.

Besides paper, a range of other materials were in constant shortage; in 1970, Soviet industry produced eight times less plywood and two times less fiber desks than did the United States.[9] The same problem applied to the making of packaging, in particular that intended for the food industry. In the West, new forms of packaging were seen as characteristic of modern life, reflective of hygienic and civilized practices. Soviet specialists recognized that the economy would significantly benefit from the availability of cheap and firm packages, and often identified food and other types of packaging as progressive materials. In the 1960s, sanitary towels and other everyday paper products were also seen as modern materials to ease female labor and the life of young mothers. But even by the end of the Soviet epoch, industry could not satisfy consumer demands for these goods. Progressivism in experimentation and professional visions ran up against a scarcity of technological infrastructure, constituting a complicated picture of socialist industrial development and the partiality of Soviet technological modernity. Even at the end of the Soviet regime, few paper boxes were used. A lot of cargo was packaged in wooden bins, which were heavier and more complicated for loading and transporting, requiring more wood in the process. Bread, meanwhile, was usually sold without packaging.[10]

The low level of manufacturing wood-based products nonetheless did not undermine the growth of production and the fact that specialists cared about wood as a rapidly shrinking resource. First, the production of new materials indeed increased compared with the pre-1945 levels, provoking alarmism over shrinking wood supplies. For example, the production of paper before and after the Second World War increased several times, but was nevertheless still not high; in 1940, 2,300 tons of paper was produced per day; 6,400 in 1960; 11,500 in 1970; 14,400 in 1980; and in 1986, 17,600 tons. Yet the volumes of round timber exported abroad were always high. In 1950, the Soviet industry exported 4.4 million tons of round wood; in 1960, it rose more than fourfold; and by 1985, 18.1 million tons were being exported.[11]

Many specialists argued that a higher level of mechanization and automation of industrial operations in forests and the manufacturing of wood-derived goods served as an important prerequisite for making the industry more efficient. Such modernization could help produce more consumer goods while leaving behind less waste from processing wood, thereby reaching the key goal of Soviet industry to produce more goods at lower costs. Industry would cope with the wood crisis by launching better mechanized and automated operations to minimize the loss of natural resources.

A LEAP TO THE FUTURE

In his memoir, *A Russian Journal*, US writer John Steinbeck wrote of his travels to the USSR in 1947. "They [Russians] love automatic machinery," he commented, "and it is their dream to be completely mechanized in practically all of their techniques."[12] In actuality, the total percentage of mechanization and automation in Soviet industry at the time was pretty low; for example, in traditional industries such as forestry, the level of mechanization of wood harvesting in 1948 was 12 percent, while the loading of wood was almost entirely done manually.[13] In 1962, of fourteen operations in forests, eleven required enormous muscle effort from workers. This not only displayed backward technology but also resulted in higher risks to the health of workers in forests undertaking this dangerous work.[14]

At that time, Soviet specialists clearly recognized that the routine mechanization of works in forests and at factories had significantly increased in developed countries over the last decades. The reference to mechanization as a superpower that could help improve economic developments appeared more and more frequently in Soviet professional and official sources. The Soviet leadership aimed to be part of global trends in automated processes and resolved to follow the world's most advanced countries, primarily the United States, filtering internal development through the lens of technological competition with the West. US technological success in automating industrial operations, particularly introducing automated assembly lines for mass production at Ford automobile factories in the early twentieth century, played a crucial role in inspiring Soviet industrialization. As early as the first decades following the revolution, the Bolsheviks transferred practices of mechanized operations and automated production inspired by Fordism and Taylorism, trying to adapt techniques like the scientific management of labor to the socialist system. Automated assembly lines, as with many other technological achievements of the US economy, were to produce capitalist fruits on socialist ground and contribute to the building of the material basis of Communism to bring the socialist society into the future. In 1948, Swiss historian Siegfried Giedion published *Mechanization Takes Command* in which he discussed the effects of mechanization on human life, confirming that mechanization had become part and parcel of US and other Western societies. Resonating with Giedion's thesis, one Soviet commentator held that in the United States, mechanization was woven into "the pattern of thought and customs," and significantly transformed the society.[15] In the USSR, full mechanization of the forestry industry was not yet a matter of reality but instead inhabited the professional imagination about future economic and societal development. In this, socialist society was to be part of the global industrialized and technologically equipped society.

While the mechanization levels of industrial operations were still low after the Second World War, specialists increasingly spoke of automation as a leap forward, bounding over the persistent realities of incomplete mechanization. The Soviet type of modernization, as imagined by specialists, was a nonlinear process, in some ways in opposition to the Marxian vision of progress; while mechanization was not yet completed, specialists already saw automation as the means of catching up with the West. As one specialist said, "If we look at US [professional] literature, we will see that all the chemical journals publish research on the use of calculation machines and automation." He insisted that specialists could really be helped not only by mechanisms for heavy forestry operations but also by calculating machines in managing the technological process in the industry and forests.[16] Many specialists connected automation and technological progress in seeking to increase the amounts of manufactured products and provide a sustainable economic production—while at the same time producing less waste. The central level went in accordance with this expert opinion: the party program of 1961, for example, spoke of the automation of industrial processes as a necessary condition for the transition to the Communist mode of labor. The regime hoped that automation as a step above mechanization would help not just to substitute human muscle with machine power, resulting in higher outputs of cut wood, less wasteful cutting, and more efficient wood processing; it would move the worker away from direct involvement in the production process and to a position of control over automated systems that would take care of production instead.

Automation in the forestry sector, including wood harvesting in forests and wood processing at enterprises, formed part of a much broader history of Soviet computers called electronic calculating machines. In the 1950s, Soviet scientists held impressive positions in cybernetics and developed several research centers that formulated promising solutions about ways to employ computers in economic activities. In the 1950s and 1960s, there were several large-scale projects to create an automated calculation system for the Soviet planned economy. Among them were Anatolyi Kitov's Economic Automatic Management System, Aleksandr Kharkevich's Unified Communication System, and N. I. Kovalev's rational system of economic control. From 1962 until 1970, Viktor Glushkov was the primary architect for the most well-known Soviet project, the National Automated System for Computation and Information Processing (OGAS).

The network was to be built with one main computing center in Moscow that regulated up to thirty local computing centers in city sites of "information flow concentrations," and an unspecified number of regional calculating centers and points of information gathering. OGAS was expected to build "electronic socialism" and network the command economy. Calculating machines were to accumulate and analyze numbers as opposed to people. In this way, automation was expected to offer a path to a technologically progressive future in which the effectiveness of economic production was vastly increased.

Automated technology fostered the techno-optimistic side of socialism and to a large extent was the foundation for the reforms of industrial management launched in the 1960s. The so-called Kosygin economic reform aimed to increase the productivity of industrial enterprises and quality of manufactured products.[17] Starting in the 1960s, many called for the introduction of calculating machines to wood-processing and other industrial operations. The meeting held in 1966 by the Central Management of the Scientific-Technical Society of the Forestry Industry was the first to state the importance of mathematical methods in pulp, paper, and wood-processing industries. They argued in favor of using computers for planning new enterprises and advised on their locations in the country based on scientific methods.[18]

Electronic calculating machines were recommended for planning and intended for the purposes of economic management. As forestry specialist D. I. Teterin wrote, "Calculating machines and mathematics provide excellent opportunities that make automation a new miracle of scientific-technical revolution.... [I]n ten to fifteen years the automated systems will become an essential part of each enterprise like the mechanization of manual labor."[19] From his point of view, this was a crucial step in building a future shorn of muscle labor. It had two other implications for forestry. First, automation promised an important means for technological control over the use of natural resources in forests. Second, automation offered the potential of both technological autonomy and speed: it could quickly transmit data from an enterprise to the central institution, thereby making forest management more effective. In fact, in this sense computers and automated systems corresponded to the will of the Soviet regime to foster technological progress, where technology was seen as a trigger, the subject of admiration, and a sign of progress. It was part and parcel of Soviet optimistic views of the future where positive human-machine relations would form the bedrock of the new society. The project of the aforementioned program of the Soviet Communist Party that began in 1961 specifically declared that the future would herald the merging of mental and physical labor in the production process. As specialist A. Mnushkin wrote in 1961, "On their cultural and technical level, workers of physical labor would reach the level of people working in the intellectual sphere." Automation would transform the worker from a cog within the production machine into an agent who controlled the technological process and thus could better control the use of nature. It would erase the difference between the worker, engineer, and scientific researchers, thereby merging creativity and physical work.[20]

In the Soviet narrative, automation—both in terms of mechanisms and the computer-led management of technological processes—appealed to Karl Marx's idea of freeing the worker from heavy and unhappy labor while offering the possibility to enjoy the results of creative work. In Marxism, the routine of industrial operations at capitalist enterprises was considered in negative terms: conveyors made workers constantly fulfill the same operations and did not allow them to think creatively. Specialists conceived of capitalist forms of automation in negative terms as well: machines in the service of capitalist purposes produced surplus, but did not free the worker to enjoy the process. Mechanized machines that excluded human force became an independent entity and liberated the person from routine labor due to the self-controlling nature of technological processes. Sophisticated automation, as specialists interpreted it, would make the human the observer and regulator—someone who "stands nearby the industrial process instead of being its major agent." It contradicted conveyor work with its scientific management of labor in which humans became "levers in the machines."[21] Now, the operations were made by machines, and the human was a manager of automated processes, giving machines the right to properly deal with nature and harvest wood.

Forestry specialists translated this idea into their industry. One example can be found in a book stating that due to automation, "the [forestry industry] personnel will be liberated from exhausting routine work and pay more attention to creative work."[22] They saw automated machines as liberators in forestry, one of the physically hardest industries. Due to more complicated professional structures than existed before, specialists proposed to use technologies to free up the labor of not only workers but also managers. In addition, some projects proposed that computers could decrease bureaucracy and increase the efficiency of management in the forestry industry. Economist N. A. Medvedev wrote in 1973 that each industrial enterprise issued about a hundred reports on annual activities and submitted numerous routine documents to the central ministry, while research and project institutions issued eighty reports per year to the ministry.[23] It required a lot of negotiations, involving personal contacts and corruption deals as well as the involvement of personal networks. Emphasizing automation and the electronic circulation of reports, he concluded that "it is not possible to survive without automation today" because "information is the nervous system of high level"; automation could make reporting and calculation independent of humans as well as their personal interests and mistakes.[24] In forestry, sixteen to eighteen new types of forestry machines were tested annually, and it was believed that three hundred employees who processed data manually could be released from calculation through automation. Specialist D. I. Teterin predicted that calculating machines in forestry would be twenty times more efficient in creating various databases recording the kinds of timber cut along with its size and other qualities, thus avoiding the mistakes, illegal operations, and irrationalities in wood harvesting.[25] Computers could project, create, and test forestry machines in different conditions using the database of natural conditions in industrial forests. Technician V. Z. Gabriel argued in 1986 that the automated system must supplement the traditional procedure of making engineering decisions. He insisted that in these new times, the potential between the engineer and machine had to be realized to make critical decisions much quicker.[26] In forestry, managers wanted to merge the economic calculation and nature to use information technology for calculating and controlling various stages of exploiting nature, or more specifically, control and manage the exploitation of forest resources.

In the 1980s, the state's faith in the power of cybernetics fizzled, and the government froze the funding and support of some initiatives. There was a concern over whether computers, with their power to transmit and collect information, would break many interactions and steady connections between institutions and individual actors. Even though ideologically the OGAS echoed the Marxist idea about making labor free and creative, it challenged the whole principle of the Soviet planned system, which was heavily based on personal connections and trust. At that time, Soviet computing obviously lagged behind the West. In the 1980s, the OGAS did not have the level of state support it had once enjoyed and became increasingly relegated to the past rather than celebrated as the future. Yet the idea of automation was still firmly connected to notions of progress. In the 1980s, the Soviet Union again planned an all-state program to develop the effective use of calculating machines and automated systems, setting a target for the full automation of all processes by the year 2000. During the whole period, sources replicated this idea: automation was something progressive and extremely important for productive development, but it was still ahead. This technical future seemed as elusive in the 1980s as it did in the 1960s.

Overall, forestry industry authorities and specialists who worked in the sector tried to justify the incorporation of mathematical approaches to production and management in forestry. They aimed to transfer automated systems to the forestry sector believing that it would offer huge boosts in controlling enterprises and the whole sector, increasing the effectiveness of resource use. The automated system was seen as having the potential to break institutional barriers by making the planned economy more workable, optimizing timber harvesting and use, and collecting data from the testing of new techniques. Being a product of rationality, a concept so beloved by Soviet specialists, the automated system was, however, a failed leap for rationalization in wood harvesting and industrial production.

BETWEEN THE MODERNIZATION AND DEMODERNIZATION OF FORESTS

Different Soviet sources evaluated the practical success of mechanization and automation in the forestry industry differently. In 1953, forestry officials wrote that the "mechanization and automation of production became widely used. The Soviet Union is the best-equipped state in the world."[27] While some individual enterprises were indeed well mechanized, other sources insisted that these claims were clearly exaggerated with respect to the whole industry; in some forestry operations, automation was not at all well developed even in later decades. The differing levels of automation broke the homogeneity of the planned economy; despite inclusion in the centralized system, each enterprise and logging company advanced separate attempts to mechanize and automate with different degrees of success. Because of interenterprise barriers, the experiments in automation undertaken by different enterprises were sometimes developed as if there was no other experience in the country. The central economy that distributed resources to the enterprises did not manage to induce the hospitable conditions required for each to choose the most promising path for automation and introduction of the most advanced technologies. The industry was characterized by enterprises that both lead and lagged behind—an unevenness that stemmed from the decisions of the center, which were based on how strategic and promising particular enterprises seemed in the eyes of state officials. At the same time, in general, the late 1950s represented a turning point for the forestry industry because both the political center and specialists admitted that in the Soviet Union, the country of "endless" forests, the forestry industry lagged behind the capitalist world. This was the time when two truths met: first, that many new technologies were in operation in other countries, demonstrating the point on the progress timeline where the Soviet forestry industry fell; and second, that introducing them was much too complicated for the planned economy, despite the constant state struggle for global technological leadership.

This also manifested in a conflict between professional visions and infrastructural realities: a desire to leap into an automated future faced practical obstacles caused by the existing infrastructures of the planned economy. Real technological possibilities and limits constrained the dream of automation. In practice, automation at factory operations and wood-harvesting sites did not represent a smooth digital move to Communism. Technological deficiency complicated industrial development, which was one of the central stakes of the Soviet regime. In the 1940s and even 1950s, the lack of mechanization at heavy operations presented a critical problem: the human remained the main lever, the primary cog. Primitive techniques based on human muscle remained pivotal. At some forestry enterprises, for instance, there remained in service a host of loader people. Engineer Sokrat Punegov described how even at a new Kamennogorsk papermaking factory built in 1949, there was no mechanization of papermaking operations in later decades. Many enterprises, and especially those that functioned in factory-based settlements, organized training for specialists on automation and control in local colleges. For enterprises, this was the only way to acquire qualified specialists as the labor turnover at Soviet enterprises was high. Sometimes up to half of the workers changed workplaces at an enterprise during the year, partly because of poor working and living conditions, and partly because forest labor remained largely seasonal. Punegov was disparaging of the level of those he called "homemade" specialists in factory-based monoindustrial towns, yet also criticized those who graduated from industrial training and industrial schools in larger cities; while they were better trained, they did not want to move to work and live in small monoindustrial settlements.[28] In forests, workloads were heavy and remained largely based on manual labor, even though new techniques were introduced such as tractors, saws, and logging trucks. At the same time, sources are replete with stories that indicate it was typical for Soviet enterprises to be equipped with outdated machines, and many struggled to acquire new techniques. Additionally, new automated and semiautomated lines frequently did not have repair details; in 1963, for example, a line in the Ural'sk logging company did not work for two months because of a broken pump and the absence of a spare.[29]

There was another contradiction in the Soviet type of modernization that lay between the new techniques and lack of skilled labor. There were different opinions about what was to be the central force of progress: machines or humans. On the local level, enterprises suffering from shortages of resources were often more interested in employing excessive numbers of workers than investing in technological reequipment. At the Chepetsk logging company in the 1950s, for instance, an enterprise characterized by low levels of mechanization, thirty-seven workers continued to load timber even as mechanization could reduce that number to twelve workers. Frequently during winter, the company stopped production and used all workers to clean snow and repair mechanisms in place of their direct production responsibilities.[30] While equipping enterprises with new techniques was a problem of the shortage economy, the enterprises were in many senses addicted to using larger numbers of workers than needed so as to always have sufficient quantities of labor. For many producers, human labor was more reliable than techniques: it was possible to use the utmost economic potential of human labor.

Modernization was in fact slow in Soviet industry and often described not as a matter of the present but instead a prospect of the future: it was through the modernization of techniques and technologies that the modern future was to be arrived at faster.[31] When one commentator wrote, "We have so much to do in the future," he meant that specialists should invest more in technologies that were weak in Soviet industry.[32] Supporting the modernization solution for creating a sustainable system of wood consumption, many specialists nevertheless still believed that humans should be assisted, and accused engineers and workers of indifference toward new techniques. The aforementioned Punegov emotionally portrayed new Soviet paper machineries as "virgin creatures" and Soviet workers as rude abusers. He complained of the rudeness of workers when they were dealing with new techniques: "Each machine can be *abused* by unskilled and unprepared cadres, and then it will not compensate the costs even though having significant economic potential."[33] Even as it appeared brash, similar criticism could be found in other publications. A lack of labor was not, as such, the most determinant problem for the Soviet economy but rather the lack of qualified and stable labor. Engineer V. Zelenin spoke in 1963 about the Bisert' logging company of the Sverdlovsk region where the first Soviet semiautomated line for cutting, sorting, and loading timber was set up. He described these lines as sophisticated machines that should be treated carefully by well-qualified and specially trained specialists. But these lines were too slowly used for full capacity, he said, because they are serviced by "sporadic people" who had only received cursory training.[34]

Technology was a means to transform raw materials into ready-made products, and Soviet producers wanted to derive as much benefit as they could from it. The principle contained in the phrase "we have to acquire total efficiency from the new techniques" (*Ot novoi tekhniki—polnuyu otdachu*) was widely disseminated in the Soviet Union. Workers often overexploited the techniques, meaning that they used machines at the highest-possible speed and capacity to produce the required amounts of products. The other side of the coin, however, was the underexploitation of techniques. For example, specialists complained that automobiles and tractors, the main productive force in logging companies, were used for only 50 to 60 percent of their total capacity because they were outdated.[35] Techniques were expected to offer everything they could to most effectively use the forest, and through this sacrifice, to realize Communism. Interestingly, the state tried to regulate this by giving prizes for a solicitous attitude to techniques. There was a tension between those specialists who thought that new techniques were required for future progress, and those who were suppressed by the plan and did not invest in learning how to work with new and more sophisticated techniques. These two poles of interaction with techniques showed how modernization and tradition merged in forestry, bringing together progressive thinking about technology, on the one hand, and old-fashioned practices, on the other. Factory newspapers often stressed that workers were unwilling to invest in self-education, but poor living and working conditions in many places of the forestry industry, high rates of labor turnover, and the suppression of the plan explained the reasons for this attitude. Progressivists complained that "introducing new techniques is slow because of egoistic love of old techniques. We must look progressively; if the old techniques are backward, we have to throw them away and introduce new techniques."[36] This simple logic, presented at the party meeting of the logging trust Sevzaples in 1955, was important: it was not unusual for Soviet production to run on old machines. An engineer of the Svetogorsk pulp and papermaking plant, I. Plakhov complained of the "firm opinion that we should not introduce new techniques [in the industry]." He specified that two production units of the same enterprise had polar attitudes toward the new techniques.[37] This seems to have been rather a personal choice, but it shows that the technological drive of the state did not always find unanimous support at the bottom level.

It is probable that some feared that the machine would come to substitute the human and therefore they continued relying on human muscles in forests and at industrial enterprises. Even in the age of rapid technological changes and dreams, in which human-machine relations were a driving force for the management and development of the forestry industry, human factors remained the priority when looking at the level of enterprises. As some specialists said, "We fulfilled the plan, taking huge pains" (*prilozhiv mnogo usiliy, plan vypolnili*). Others stressed that "not banner resolutions and declarations but cadres, humans, define the developments of the techniques."[38] If cybernetics relied on equality in human-machine interactions and freeing labor from routine work, forestry workers and engineers (with the exception of those I identify as progressivists who proposed the modernization solution to the wood crisis) still saw the human factor as decisive. L. Ross, the head of the Technical Management of the Forestry Ministry, observed once that "our engineers and technicians are not worse than Canadian, Swedish, and French. And it is a pity that overall productivity in [the West] is two to three times higher."[39] In opposition, progressivists relied on new technologies; for them, the human factor was to enhance the operations of techniques, which in turn served the Soviet economy. Yet in the 1980s, up to half of all operations in the forestry industry were still done manually.[40] In 1984, the forestry industry remained an industrial branch where the level of mechanization of forestry works was only 40 percent. Specialists referred to this fact in explaining the slow reforestation and ineffective use of forest resources.[41]

By the end of the Soviet period, modernization did not transform into a full-fledged driving force for the efficient development of socialist forestry and remained a matter of progressive dreamscapes. As in the imperial and experimental solutions, implementing mechanization and automation required not only professional approval but also state action, political will, and serious financial investments. The role of the state was ambiguous, though: while insisting on the importance of technological improvement, it did not fully complete mechanization and automation both because of a lack of funding and the resistance that emerged at some enterprises that emphasized the dominance of humans. The planned economy thus developed a complicated environment where successes in the mechanization of forest work and some industrial processes at particular enterprises coexisted with the continuation of manual labor. Both coevolved with the progressivist picture of the future. At the same time, attempts to improve technology for better exploiting forests were sparked by the desire among specialists to decrease the amount of forestry waste and wasteful cutting, and improve the inefficient production chain. Machines and automated equipment were expected to run forests with greater precision and accuracy than humans could, and sought to make workers and specialists supervisors of industrial processes. This contributed to industrially embedded ecology, showing how progressive technologies were to participate in saving forests due to a more careful and accurate approach to them. As a space of modernizing and demodernizing practices, the forestry industry gave room for thinking about progressive technologies as more efficient tools for keeping forests sustainable, even as many of these were not effectively implemented.

EPILOGUE: THE CONTINUITY AND DISRUPTION OF GREEN INDUSTRY IN THE (POST)-SOVIET ERA

In summer 2021, my father and I traveled to the village of Miynala in Russian Karelia near the Russian-Finnish border. This tiny place, formerly part of Finnish territory, found itself on the borders of the USSR after the Soviet-Finnish War of 1941–1944. My father had lived there during his childhood from the late 1950s on, and had left the place in his twenties with memories of it as a land surrounded by towering evergreen fir trees. Returning after twenty-five years, he regretfully found that some parts of the previously dense fir forest near his childhood home had been clear-cut. His tears and long silence after what he saw spoke of his pain; an open space where evergreen fir trees once stood bore visual testimony to the realities of tree cutting, transforming the cut tree from an industrial material to a symbolic object of memory. It was not a total shock, however; living in Karelia during the 1990s and 2000s, I myself used to see logging trucks with full loads of cut timber headed to neighboring Finland for export.

The post-Soviet decades have a great deal in common with the practices of Soviet planned economy in terms of wood harvesting and the technological levels of the forestry industry. Rapid deforestation continued, though frequently through more illicit channels than before, with numerous companies cutting and selling huge amounts of timber abroad. In the early 1990s, official statistics recorded decreases in the overall volumes of cutting compared to the figures of the late perestroika era, but they were nonetheless intensive. Cutting processes continued developing in the eastern parts of Russia where some infrastructure and enterprises had been established by state socialism. Russian domestic demand for wood had been and still remains small, and the amounts of cut wood significantly exceed the consumption capability of the inner market. The predictions made by Soviet specialists about the rapid and intensive growth of industrial production due to sophisticated technologies have not been realized. Instead, the volumes of harvested wood have increased, now cut mainly by private companies and leading to the wood crisis. In the 1990s, private profit became the most important category and motivation for harvesting, and to a large extent maintained the intensive cutting of forests in Russia. The fate of the wood-harvesting industry resonates with that of other extractive economies that export wood instead of developing full processing circuits inside the country.

Like coal, oil, and gas, wood has remained an important raw material for the resource-based economy, maintaining its dependence on natural riches. And in the 1990s, as previously, Russian specialists who remained employed in the forestry industry often described the situation prevailing in wood harvesting and processing as difficult and careless (*beskhozyaistvennyi*). From a professional point of view, the wood crisis persisted, and remained a real danger because of the ongoing and extensive exploitation of forests. Public attitudes toward deforestation, especially of the illegal kind, were also critical toward cuttings and illegal exports of wood.

Many post-Soviet forestry practices have remained as wasteful as they were in the Soviet Union, and post-Soviet industry is still largely based on old Soviet technical infrastructures—even as the immediate post-Soviet decades proceeded with disavowals of prior Soviet experience in forestry, representing a rupture between past and present. The post-Soviet industry has refused to forge a postsocialist forestry industry based on past models, condemning them as useless Communist experiments. It has also relegated to the past the socialist attempts to predict and explain the wood crisis along with those pioneering projects of rational resource use and no-waste production. This break with socialism incinerated the forms of industrially embedded ecology that had emerged over the course of the last decades, even as rapid deforestation, the wasting of forests, and pollution continued to be notorious realities in Russia. The numerous professional proposals, experiments in no-waste production, and ideas around increasing the efficiency of manufacturing from wood that developed as part of Soviet industrially embedded ecology were attributed to a bygone ideology, thrown into the "dustbin of history." Many engineers again emphasized the importance of learning from Western experience, referring, for instance, to the so-called Scandinavian type of cutting (*skandinavskaya rubka*)—sometimes also identified as Finnish—or a type of cutting techniques employed in northern Europe (primarily Finland and Sweden).[1] Russian foresters have, as in the past, connected these models to efficiency and nondepletion technologies, underscoring the importance of technology transfer from the West. Yet they do not consider the lessons learned through previous Soviet experience.[2] This has partly echoed the ways in which the Soviet industry denied the industrial know-how developed over the czarist period, like the experiences of manufacturing from reed. The post-1945 period did not invent but rather developed previously invented ideas, such as the complex and rational use of forest resources found in the 1930s at least. Equipped with better technologies and more environmentalist attitudes, however, it saw czarism and often the pre–Second World War period as backward and hindering of modern development, while explaining earlier experiments as pioneering and useful yet technologically lagging behind. Similarly, postsocialist Russian forestry has not seen the Soviet past as something to positively reflect on. Cyclic denial of industrial experience has derived from the neglect of the preceding political and ideological system, which has always been relegated to a less developed and therefore useless stage in the linear timeline of history.

Neglecting Soviet practices of industry-nature relations serves as an important indicator of the low priority given to nature in modern Russia in general. While for a period of Vladimir Putin's presidency, increasing state interest toward ecology and the environment could be detected, it did not last long. In 2017, the Russian government declared a "year of ecology" in the country to increase public and professional attention on environmental problems, expressing at least a formal state concern over the environment. In just a few years, however, political transformations, most starkly expressed in the war actions that the Russian government launched in Ukraine in February 2022 and in declaring a number of environmental organizations as "foreign agents," put nature in Russia on the back burner.[3] Climate change, rapid deforestation, and large-scale forest fires remain pivotal issues, but they are on hold as themes of secondary importance. A radical change in the state's approach to the environment over the last few years has put political pressure on environmental activists, despite environmentalist declarations made by the state.

The post-Soviet era has seen the continued and extensive use of naturally growing forests and witnessed only a small share of reforestation. Russia has also continued the export of round wood and not invested a great deal in the already weakly invested wood-processing industry. As in the Soviet Union, Russia still consumes significantly less paper per person than developed liberal economies.[4] Along with this, it suffers from numerous environmental problems related to forest depletion and pollution in its different regions. Local dwellers of Svetogorsk, for instance, a town on the border with Finland where a huge pulp and papermaking plant operates, describe colored snow and dirty air as familiar features of their urban landscapes.[5] Huge swathes of forest in Siberia and the Far East have been cut by Russian and foreign companies, particularly Chinese ones. In 2017, Russian sawmill products made up 30 percent of all sawmill products exported to China, according to some calculations.[6] Forests wasted by harvesters and local dwellers, along with incredible forest fires, are still major problems in Russian forests to this day, especially those in Siberia. The processing of wood waste remains a vague prospect for the future due to little interest and poor forest infrastructure.[7] In 2022, the state issued a complete ban on exporting round timber from

Figure E.1 Clear-cut of forest in the Russian north. *Source:* "Ostanemsya bez lesa," https://arh.aif.ru/society/people/1307897.

Russia, exemplifying not only the break of economic operations between Russia and the West but also showing how deep the problem of wood depletion is in Russia.

The materiality of wood nevertheless remains important as political crisis and economic sanctions on Russia imposed by Western countries since 2014 have not radically diminished consumer demands for paper, furniture, and other wood-based products. In 2022, economic sanctions were sharpened, interrupting supplies of the chemicals and components required for the high processing of wood. The price of printing paper in Russia has especially skyrocketed. Some papermaking enterprises were even temporarily shuttered because of the lack of chemicals, revealing the backwardness of the papermaking industry. While there was still raw wood available, enterprises lacked materials needed for manufacturing and bleaching pulp, with most chemicals still imported from abroad and blocked by Western economic sanctions. This revealed a strong compensational function of Western imports to the Russian economy (often defined as "dependence" in Russian media and political circles) and the incapability of the latter to immediately adjust to the extreme conditions. Interestingly, in the wake of the political and economic crisis, several Russian newspapers wrote about a technological innovation developed recently by Russian scientists to process cow parsnip into bleached cellulose.[8] Previously seen as a noxious weed that presented health hazards, cow parsnip, which grows across Russia, was rethought as a useful industrial material. Yet the Soviet projects of the past that sought to recycle annual plants were entirely forgotten in the enthusiasm for cow parsnip.

Soviet experience in this sense provides lessons for contemporary Russia. But it also offers insights for liberal societies in terms of proposing theory and experiments that led to specific types of industry-nature interactions. Based on rationality and the complex use of natural resources, the Soviet approach sought to achieve no-waste production and sustainable economic growth with a stable supply of material goods. Considering the Soviet experience of developing solutions to prevent a wood crisis requires drawing a distinction between professional expectations and experimenting, on the one hand, and implementation, on the other. This is an important difference, emphasizing that experimentation and the professional dreamscapes that revolved around the concepts of progress and technological advancement met the materiality of socialism and found numerous infrastructural obstacles. In the twentieth century, the forestry industry was not simply a critical industrial branch but significantly depended on state investments and material support too. Natural resources played a crucial role in this process: they were materials, both physical and symbolic, constituting the surrounding environment for the people. Wood provided the green power to both build the industrialized society and uphold narratives of Soviet might, especially in the context of competition with the West. From the 1950s to the 1980s, Soviet specialists proposed to restructure the raw material base in order to realize effective industrial development, prioritizing new spheres of consumption while seeking to manufacture as much as possible with minimal costs. Initiatives moved by alarmism usually were supported by state political leaders in word, but rhetoric and partial funding met serious obstacles, with militarism—an obvious priority under socialism—drawing incredible amounts of funding away from these efforts. In the Soviet Union, for many forestry specialists, making cardboard packages for food was a symbol of progress and inevitable consumerism, while producing arms remained a greater priority for the state. Consumer paper was the subject of rhetoric about a future Communist society and often was not abundantly available to Soviet citizens. In the industry, a huge gap between the harvested volumes of wood and real production of consumer goods therefore was a pivotal issue during the whole Soviet period and afterward. As one book put it in the year of the demise of the Soviet state, "The main quantity of sawmilling materials is consumed in Canada, Sweden, Norway, and Finland. We, being the forest superpower, should obviously aim at the same level."[9] This quote perfectly reflects the gap between the availability of natural resources and technological possibilities, attributing forest to *superpowerness*, but at the same time stressing that the Soviet Union was *behind* developed forested economies. Even though the forestry industry was an important supplier of a vast range of products, its share in the industrial sector was small—only 5 percent of the whole production in 1988.[10] The ambitious thinking about nature-given resources as evidence of resourceful might and superiority still coexists with poor technological infrastructures and a lack of sophisticated technologies in present-day Russia.

Importantly, though, forest alarmism among Soviet specialists after the Second World War had challenged Russia's traditional *resourceful imperialism* long grounded in history along with its slow colonization of densely forested and scarcely populated lands from the Urals eastward. The professional investigation of forests explained the clear difference between wood as an item of economic prosperity versus cultural myth. Huge forest covers of the eastern lands of Siberia and the Far East would serve the economy, but the economic abundance of wood, specialists explained, was more public illusion than economic reality. Colonial advancement to green unspoiled lands in the eastern parts of the USSR demonstrated the tension between mythmaking around the forest as a cultural actor and the industrial application of wood, showing that the depletion of industrial wood was a real danger to sustainable economic growth. Even so, extensive advancement into the eastern parts of the country to technologically colonize unspoiled forests was a project grounded in rationality as well as a deep expectation of turning a new page and building new organizational forms in the forestry industry there. Industrialism and a passion for technocracy instigated by the availability of scientific and technological instruments to transform wood, waste, and annual plants into sophisticated consumer materials, along with a strong belief in the power of progress, provoked concerns about the future "fuel" for economic development. Technology acted in this sense not as a risk for nature but instead as a possibility to reconcile economic growth and nature together as two ingredients of the modern world. Increasing the volumes of production was important for Soviet specialists as it could change society, bringing technological achievements into social life and bringing forth a diversity of material goods—one indicator of the quality of life. Despite the fact that public views still often replicated the image of forests as historically bestowed abundance, professional alarmism appealed to the scarcity of industrial wood. In contemporary Russian society, forests are still seen as vast green riches, and ongoing overexploitation is leading to depletion. This highlights the continuity between Soviet professional predictions of the past and the likely future of Russian forests: the Soviet experience had demonstrated the fragility of Russian forest imperialism and the scarcity that belied the apparent abundance of wood stocks.

All three solutions for preventing a wood crisis examined in this book—imperial, experimental, and modernization—implied the more rational use of wood to facilitate more sustainable economic growth. Specialists remained industrialists and never truly became ecologists. Production as a paradigm was never criticized by the Soviet power and professional communities; rather, it was the conventional basis for the economy of a society impassioned by overindustrialization. Ideologically, it was crucial for completing the project of forging the material basis to build a Communist society. Natural resources played a crucial role in this process.

Industrially embedded ecology was found in all of these solutions, and derived from an economic purpose to increase production through more efficient methods explained as rational and waste minimizing. A desire to save on costs and improve a wasteful economy, as specialists recognized it, provoked a desire to decrease the burden on forests by different means. For the colonial model, moving harvesting and wood-processing capacities to the "underexploited" eastern forests could work to decrease the volume of harvesting in the exhausted northwest. Extensive advancement to the east thus aimed to save old industrial regions from total depletion and a future of desertification. Those who developed experiments in waste recycling and reed processing did so in the hopes of rendering the use of cut wood more efficient, making cutting both less wasteful and less necessary in providing the material required for industrial production. The modernization model relied on intensifying technological processes as well as increasing the levels of mechanization and automation to prevent wood loss during harvesting and wood processing at industrial enterprises. Effectiveness implied minimal loss, and stimulated numerous initiatives to save costs and reduce wasting from woodcutting operations. These models were all based on a technocratic vision of nature and relied on a few important notions that circulated in the Soviet economy: rationality, saving or economizing, and the complex use of natural resources. Rationality referred to minimizing waste, emphasizing that all possible resources should be used in industrial processes. The complex use of natural resources emerged as an influential principle in late Soviet industrial development. Complexity could help maintain sustainable resource supplies and industrial growth; now, not only the best wood could be used, but all parts of trees that had hitherto been left as waste accrued industrial value.

Attempts to make progress in forest exploitation failed in practice, however. This book has told the story of how a huge country that tried to deal with its own natural riches did not manage to *implement* professional solutions to provide sustainable development. The failure of the planned economy resulted in the incapacity to implement expert ideas even as the central state made investments in new industrial enterprises and infrastructures. Failure is, as such, woven through the chapters of this book to demonstrate that state socialism and central planning produced progressive views of the interplay between industry and nature, but simultaneously set obstacles in the path of implementing them. Innovative ideas expressed by specialists required modern infrastructures and long-term investments, but were often not met with sufficient resources. As in the case of waste processing and the use of annual plants, material shortages and the lack of expensive infrastructure created obstacles for innovation and production.

Yet attributions of failure should not be made wholesale. None of the proposed solutions to the wood crisis succeeded at full capacity, yet *neither* were they all total failures in terms of their discursive value and experiment. Found in backward material infrastructure, projects designed for saving wood from depletion in order to make more diverse goods for consumption instigated discussions and experiments to make industry greener under state socialism. Compared with previous decades, many ideas derived from late Soviet projects were based on the belief in more progressive technologies and the need for sustainable consumer production to address earlier concerns about forests. By the mid- to late 1970s, specialists developed even more sustainable industrial thinking and spoke of the ecological as well as recreational functions of forests, insisting on the importance of keeping them safe by deploying new methods for acquiring raw materials in place of brutal clear-cutting. They did not talk about the conservation of forests as such but instead sought out solutions beyond legislation to most efficiently use wood and wood-related materials. This brought greater environmentalism to industry; using waste material was important because of the envisaged lack of wood, but also because of the noneconomic functions of the forest. Undoubtedly, this view on industrial sustainability developed in the context of increasing environmentalism in the Soviet Union and beyond, and reflected transformations in the views of specialists working in the forestry industry. Using waste would keep forests clean, decrease the number of forest fires, and allow for the manufacture of numerous modern consumer products.

Importantly, the perestroika period again surfaced a gap between the professional dreamscape and realities of implementation, and stimulated professional discussions about the need for making the forestry industry more effective—that is, more productive with lesser costs—and the use of natural resources more rational. Transforming industrially embedded ecology from a set of discourses to programs of action was still a prioritized aim of late Soviet producers. Recognizing the failures of all three models by this time, specialists continued emphasizing that the use of waste, reed recycling, and other technologies was urgent—in ways that were similar to the tenor of their discussions decades before. Likewise, they spoke of the need to develop forest settlements and social infrastructures close to depleted forests, even as those forests remained a marker of social decay in the former Soviet Union.[11] Perestroika, which began from the drive for economic modernization, resulted in loud discussions about Soviet political and economic problems, but did not result in moves to modernize the industry. Once more, the political center set long-held aims for overcoming wastage and the lack of efficiency in the forestry industry, but explained them as matters for the future. What made this period different from earlier decades was a changed view of forests. By the 1980s, specialists increasingly described nature as having its own agenda as well as being important for the broader ecosystem and human life. By this time, they had also come to connect waste not only to industrial value; they tied it to an environmental practice, advancing the need to keep nature pristine even though economic development remained a priority.

In recent decades, scholars have sought to explain green strands of past and present political regimes. Some have demonstrated the evolution of green nations of capitalism, discussing how countries like Norway and Germany were much greener than previously imagined. Peder Anker, for instance, stresses how scholars and activists portrayed Norway as place of environmental stability and a pioneering green nation on the periphery of a contaminated and conflicting world.[12] Frank Oekötter argues that Germany founded a sustainable future in the late twentieth century because of a strong environmentalist initiative.[13] In contrast, East Germany specifically and every other socialist regime, Arvid Nelson concludes, destroyed the environment and to a large extent met their end because of their ignorance of ecology.[14] In these studies, capitalism is to a large degree portrayed as the cradle for green ecology in the context of the growing environmentalism of late decades, exemplifying that some liberal economies *finally* produced a strong concern for nature.[15] They debate with a robust and old (yet still powerful) discourse expressing a critical view of capitalism as depleting nature because of its egoistic interests. In particular, an iconic story about the 1930s' dust bowl is one of many examples invoked to demonstrate how concerns about quick profit can result in the devastation of pristine nature.[16]

Beyond the discussion about capitalism and nature, there is a changing view of socialism as not exclusively ecocidal but in fact containing more strands of green than previously countenanced. Zsuzsa Gille, for instance, discusses the waste regime in socialist Hungary as a model of socialist interaction with the material remains of socialist industrialization.[17] Stephen Brain insists on a more careful approach to nature under Stalinism in the USSR that saw a period of forest conservation.[18] Petr Jehlička and Joe Smith provide evidence for a postwar Czech environmentalism through nature conservation and education, which remained alive and influential after the demise of socialism there.[19] While this measurement of both the good and bad effects of human activities on nature should not be exaggerated, socialist experiments should certainly not be reduced to the ecocide model. *The Green Power of Socialism* has not sought to enter the fray to advance another example of socialism's benefits for nature.

Instead, it has shown how economic priorities and the passion of industry supported a specific, complex, and at times contradictory attitude toward nature among industrialists. The Soviet Union, a space where large-scale industrialization became a national idea, was home to a contradiction, demonstrating how industrial ecology coexisted with a wasteful economic regime. Soviet specialists raised their voices against the depletion of forests in the name of producing more industrial goods in the future and developed a discourse of industrially embedded ecology as a by-product of rapid industrialization. They argued that the transition from extractive to intensive economy was to be based on saving natural resources in the name of future economic development. The professional attitude toward forests, even as it remained critical and alarmist, underscored the liminality of forests along with their ability to undergo a substantial economic, cultural, and political transition. The forest was transformed from an abundant material resource exclusively serving humans to an actor in its own right that provided a great service for society. The forest—the green coverage—was a power that compelled socialism to rethink its relation with nature while remaining committed to industry as its conventional priority. In the context of ongoing environmental crisis in the world today, we stand to learn a great deal from this historical transformation about how to build a better dialogue with this green power for a sustainable future.

NOTES

INTRODUCTION

1. Aleksey Bukshtynov, Boris Groshev, and Georgiy Krylov, *Lesa* (Moscow: Mysl', 1981), 316.

2. Aleksey Bukshtynov, *Lesnye resursy SSSR i mira* (Moscow: Izdatel'stvo Ministerstva sel'skogo khozyaistva SSSR, 1959), 8; Prokopiy Vasil'ev, *Lesnye resursy SSSR segodnya i zavtra* (Moscow: Znanie, 1969).

3. See, for example, "Lesnye resursy Rossii," Astrakhanskaya oblastnaya nauchnaya biblioteka, accessed June 7, 2023, aonb.astranet.ru/lesnyie-resursyi-rossii-2.html.

4. O vazhnosti tsellyulozno-bumazhnoi promyshlennosti, 36 sessiya uchenogo soveta, 24 noyabrya 1959 goda, TsGANTD SPb (Central State Archive of Scientific and Technical Documentation in Saint Petersburg), F. 303, Op. 13, D. 441, L. 12.

5. Vasiliy Rubtsov and Prokopiy Vasil'ev, *Lesnoe khozyaistvo SSSR za 50 let (1917–1967)* (Moscow: Lesnaya promyshlennost', 1967), accessed November 26, 2022, https://www.booksite.ru/fulltext/za5/let/2.htm. In 1987, the forestry industry manufactured more than 3 percent of all industrial products in the Soviet Union. See Petr Zakrevskiy, *Razvitie lesozagotovitel'noi promyshlennosti za 70 let sovetskoi vlasti* (Moscow: VNIPIEIlesprom, 1987), 1.

6. See, for example, Susan E. Reid, "Communist Comfort: Socialist Modernism and the Making of Cozy Homes in the Khrushchev Era," *Gender and History* 21, no. 3 (November 2009): 465–498; Alexandra Staub, *The Routledge Companion to Modernity, Space, and*

Gender (Abingdon-on-Thames, UK: Routledge, 2020), 412; Magdalena Eriksroed-Burger, Heidi Hein-Kircher, and Julia Malitska, eds., *Consumption and Advertising in Eastern Europe and Russia in the Twentieth Century* (Cham, Switzerland: Palgrave Macmillan, 2023).

7. "Vdokhnovlenno trudit'sya nad vypolneniem resheniy XXII s'ezda KPSS," *Derevoobrabatyvayushchaya promyshlennost'*, no. 12 (December 1961): 1.

8. For more on the roles of experts in various regimes, see, for example, Loren Graham, *The Ghost of the Executed Engineer: Technology and the Fall of the Soviet Union* (Cambridge, MA: Harvard University Press, 1996); Timothy Mitchell, *Rule of Experts: Egypt, Techno-Politics, Modernity* (Berkeley: University of California Press, 2002); David Brydan, *Franco's Internationalists: Social Experts and Spain's Search for Legitimacy* (Oxford: Oxford University Press, 2019).

9. Illustrations of this include specialists' activities in environmental protection, cybernetic projects, and decision-making about industrial building. See, for example, Slava Gerovich, *From Newspeak to Cyberspeak: A History of Soviet Cybernetics* (Cambridge, MA: MIT Press, 2002); Elena Kochetkova, "An Ecological Controversy: Soviet Engineers and the Biological Treatment Method for Industrial Wastewater in the 1950s and 1960s," *Ab Imperio*, no. 1 (2019): 153–180; Roman Abramov, "Engineering Work in the Late Soviet Period: Routine, Creativity, and Project Discipline," *Sociology of Power* 32, no. 1 (2020): 179–214.

10. O vazhnosti tsellyulozno-bumazhnoi promyshlennosti, 36 sessiya uchenogo soveta, 24 noyabrya 1959 goda.

11. There is a huge and long debate on Soviet modernity. See, for example, Anna Krylova, "Soviet Modernity: Stephen Kotkin and the Bolshevik Predicament," *Contemporary European History* 23, no. 2 (May 2014): 167–192; Michael David-Fox, *Crossing Borders: Modernity, Ideology, and Culture in Soviet Russia* (Pittsburgh: University of Pittsburgh Press, 2015); Alexey Golubev, *The Things of Life: Materiality in Late Soviet Russia* (Ithaca, NY: Cornell University Press, 2020).

12. Even though I do not employ Bruno Latour's methodology, "black box" refers to his terminology. See Bruno Latour, *Science in Action: How to Follow Scientists and Engineers through Society* (Cambridge, MA: Harvard University Press, 1987), 288.

13. Robert N. North and Jan J. Solecki, "The Soviet Forest Products Industry: Its Present and Potential Exports," *Canadian Slavonic Papers* 19, no. 3 (September 1977): 281–282.

14. Brenton M. Barr and Kathleen Braden, *The Disappearing Russian Forest: A Dilemma in Soviet Resource Management* (Washington, DC: Rowman and Littlefield Inc., 1988), 252. See also Brenton Barr, "Regional Variation in Soviet Pulp and Paper Production," *Annals of the Association of American Geographers* 61, no. 1 (1971): 45–64; Brenton Barr, "The Forest Sector of the Soviet Far East: A Review and Summary," *Soviet Geography* 30, no. 4 (1989): 283–302. For a more recent geographic perspective of the environment in Russia, see Jonathan Oldfield and Denis J. B. Shaw, *The Development of Russian Environmental Thought: Scientific and Geographical Perspectives on the Natural Environment* (Abingdon-on-Thames, UK: Routledge, 2018), 212.

15. Paul Josephson, "Industrial Deserts: Industry, Science and the Destruction of Nature in the Soviet Union," *Slavonic and East European Review* 85, no. 7 (April 2007): 294. See also, for example, Paul Josephson, *Resources under Regimes: Technology, Environment, and the State* (Cambridge, MA: Harvard University Press, 2006), 278; Henry Dumon, "Ecocide in the Caspian Sea," *Nature* 377, no. 6561 (October 1995): 673–674; Murray Feshbach, *Ecocide in the USSR: Health and Nature under Siege* (New York: Basic Books, 1993), 400.

16. Arvid Nelson, *Cold War Ecology: Forests, Farms, and People in the East German Landscape, 1945–1989* (New Haven, CT: Yale University Press, 2005), 26.

17. See, for example, Cheng Li and Yanjun Liu, "Selling Forestry Revolution: The Rhetoric of Afforestation in Socialist China, 1949–61," *Environmental History* 25, no. 1 (January 2020): 62–84.

18. Ted Benton, ed., *The Greening of Marxism* (New York: Guilford Press, 1996); Douglas Weiner, *A Little Corner of Freedom: Russian Nature Protection from Stalin to Gorbachev* (Berkeley: University of California Press, 1999), 570; Nicholas B. Breyfogle, "At the Watershed: 1958 and the Beginnings of Lake Baikal Environmentalism," *Slavonic and East European Review* 93, no. 1 (January 2015): 147–180; Elena Kochetkova, "Between Water Pollution and Protection in the Soviet Union, Mid-1950s–60s: Lake Baikal and River Vuoksi," *Water History* 10, no. 2–3 (February 2018): 223–241; Arran Gare, "The Environmental Record of

the Soviet Union," *Capitalism Nature Socialism* 13, no. 3 (May 2008): 52; Stephen Brain, *Song of the Forest: Russian Forestry and Stalinist Environmentalism, 1905–1953* (Pittsburgh: University of Pittsburgh Press, 2011), 240. For a critique of Brain's book, see Nicolai Dronin, "Review of Stephen Brain, 'Song of the Forest: Russian Forestry and Stalinist Environmentalism, 1905–1953,'" *Soviet and Post-Soviet Review* 39, no. 2 (2019): 280–282.

19. For a discussion of how technocrats spoke against the devastation of natural resources, see Graham, *The Ghost of the Executed Engineer*, 154. See also Vasiliy Makurov, *Razvitie lesnoi promyshlennosti Evropeiskogo Severa SSSR v poslevoennyi period (1946–1955)* (Petrozavodsk: Kareliya, 1979), 78; Valentin Tatarinov, *Lesnoi kompleks: sostoyanie i perspektivy razvitiya* (Moscow: Lesnaya promyshlennost', 1989), 352; Nikolay Burdin, Vladimir Shlykov, V. A. Egornov, and Viktor Sakhanov, eds., *Lesopromyshlennyi kompleks: sostoyanie, problemy, perspektivy* (Moscow: MGUL, 2000), 473; Oleg Nevolin, Sergey Tret'yakov, Sergey Erdyakov, and Sergey Torkhov, eds., *Lesoustroistvo* (Arkhangelsk: Pravda Severa, 2003), 583; Ilya Shegel'man and Oleg Kulagin, "O vklade lesnogo sektora v ekonomiku SSSR v period Velikoi Otechestvennoi voiny (1941–1945 gg.)," *Nauchnyi zhurnal KubGAU* 78 (April 2012): 1–10.

20. Andy Bruno, *The Power of Nature: An Arctic Environmental History* (Cambridge: Cambridge University Press, 2016), 271.

21. Daniel Schneider, *Hybrid Nature: Sewage Treatment and the Contradictions of the Industrial Ecosystem* (Cambridge, MA: MIT Press, 2011), 222. On environmental justice that shows the merger of technology and the environment, see Gwen Ottinger and Benjamin R. Cohen, *Technoscience and Environmental Justice: Expert Cultures in a Grassroots Movement* (Cambridge, MA: MIT Press, 2011), 312. For a comprehensive review of approaches to Soviet and Russian environmental development, see Jonathan Oldfield, *Russian Nature: Exploring the Environmental Consequences of Societal Change* (Abingdon-on-Thames, UK: Routledge, 2005). See also William Boyd, *The Slain Wood on Environmental Consequences of Paper Making in the US South* (Baltimore: Johns Hopkins University Press, 2015), 376.

22. Eglė Rindzevičiūtė, *the Power of Systems: How Policy Sciences Opened up the Cold War World* (Ithaca, NY: Cornell University Press, 2016), 306.

23. Rosalind Williams, "Crisis: The Emergence of Another Hazardous Concept," *Technology and Culture* 62, no. 2 (2021): 521–546. On the official making of an ecological crisis, see Julia Lajus, "Soviet Official Critiques of the Resource Scarcity Prediction by Limits to Growth Report: The Case of Evgenii Fedorov's Ecological Crisis Rhetoric," *European Review of History: Revue européenne d'histoire* 27, no. 3 (2020): 321–341.

24. On other countries, see Lawrence Teeter, Benjamin W. Cashore, and Daowei Zhang, *Forest Policy for Private Forestry: Global and Regional Challenges* (Egham, UK: CABI), 307.

25. James C. Scott, *Seeing like a State: How Certain Schemes to Improve the Human Condition Have Failed* (New Haven, CT: Yale University Press, 1998), 4.

26. On Russian inner colonization, see Alexander Etkind, *Internal Colonization: Russia's Imperial Experience* (Cambridge, UK: Polity Press, 2011).

27. On earlier developments of the Soviet forestry industry, see Ivan Zykin, *"Zelenoe zoloto" industrializatsii: lesopromyshlennyi kompleks Ural'skogo regiona v kontse 1929 g.–pervoi polovine 1941 g* (Ekaterinburg: OOO Universal'naya Tipografiya "Al'fa Print," 2021).

28. On the Soviet planned economy, see Philip Hanson, *The Rise and Fall of the Soviet Economy: An Economic History of the USSR from 1945* (London: Langman, 2003).

CHAPTER 1

1. "Ne tol'ko rubit', no i vosstanavlivat'," *Master lesa*, no. 7 (July 1963): 1.

2. E. I. Lopukhov, "Osvaivat' i berech' bogatstva taigi," *Lesnaya promyshlennost'*, no. 6 (June 1961): 4.

3. Tikhon Petrov, *Les i ego znachenie dlya narodnogo khozyaistva SSSR* (Moscow: Lesnaya promyshlennost', 1964), 9. The author did not explain the choice of these particular countries as an illustration.

4. "Bogatstva vostochnoi Sibiri na sluzhbu narodnomu khozyaistvu," *Lesnaya promyshlennost'*, no. 7 (July 1958): 27.

5. Anatoliy Shirokov, *Dal'stroi v sotsial'no-ekonomicheskom razvitii severo-vostoka SSSR. 1930–1950-e gg* (Moscow: Rossiyskaya politicheskaya entsiklopediya, 2014), 45.

6. *Lesnoe khozyaistvo SSSR za 50 let. Gosudarstvennyi komitet lesnogo khozyaistva Soveta Ministrov SSSR* (Moscow: Lesnaya promyshlennost', 1967), 312.

7. A. A. Baitin, "Osnovy sovetskogo lesoustroistva," *Lesnaya promyshlennost'*, no. 5 (May 1947): 7.

8. N. R. Pis'mennyi, "Les—velichaishee nashe bogatstvo," in *Les—nashe bogatstvo* (Moscow: Goslesbumizdat, 1962), 16–30.

9. "Zadacha pervoocherednoi gosudarstvennoi vazhnosti," *Lesnaya promyshlennost'*, no. 8 (August 1947): 1.

10. "Bol'she vnimaniya izobretatelyam i ratsionalizatoram," *Derevoobrabatyvayushchaya promyshlennost'*, no. 5 (May 1953): 1.

11. "Delo nashei rabochei chesti," *Les—stroikam*, no. 30 (July 1979): 2.

12. N. S. Okhlopkov, "Lesa Yakutii na sluzhbu narodnomu khozyaistvu," *Lesnaya promyshlennost'*, no. 6 (June 1982): 4.

13. Pis'mennyi, "Les—velichaishee nashe bogatstvo," 16.

14. Zadachi partorganizatsii v povyshenii kachestva produktsii, 1968 g., TsGAIPD (Central State Archive of Historical and Political Documentation), F. R-1542, Op. 4, D. 166, L. 89.

15. G. Sychevskiy, "Lesnuyu otrasl' novykh rayonov—na sovremennuyu osnovu," *Lesnaya promyshlennost'*, no. 9 (September 1971): 14.

16. "Delo nashei rabochei chesti," 1.

17. Mikhail Serdyukov, *Prirodnye bogatstva Karelii na sluzhbu narodu* (Petrozavodsk: Kareliya, 1959), 4.

18. Razrabotka tekhnologii pererabotki v kormovye drozhzhi sul'fitnogo shcheloka ot varki trostnika primenitel'no k usloviyam Kzyl-Ordynskogo TsBK, 1962 g., TsGANTD SPb (Central State Archive of Scientific and Technical Documentation in Saint Petersburg), F. R-325, Op. 24, D. 1751, L. 4.

19. Leonid Leonov, *Russkiy les* (Moscow: Khudozhestvennaya literatura, 1988).

20. Spravka o rezul'tatakh proverki Postanovleniya VS KASSR ot 18.06.1961 goda ob okhrane prirody, NARK (National Archive of the Republic of Karelia), F. 3432, Op. 1, D. 242, L. 7.

21. Elena Kochetkova, "Industry and Forests: Alternative Raw Materials in the Soviet Forestry Industry from the Mid-1950s to the 1960s," *Environment and History* 24, no. 3 (2018): 323–347.

22. Mikell P. Groover, *Automation, Production Systems, and Computer-Aided Manufacturing* (New York: Pearson College Div, 2007), 815; David F. Noble, *Forces of Production: A Social History of Industrial Automation* (New York: Routledge, 2011), 444.

23. Donald Bowles, "The Logging Industry: A Backward Branch of the Soviet Economy," *American Slavic and East European Review* 17, no. 4 (December 1958): 432–433.

24. "Vpered po puti mekhanizatsii lesozagotovok," *Lesnaya promyshlennost'*, no. 8 (August 1950): 1.

25. Tsentr NII mekhanizatsii i energetiki LP SSSR (TsNIIME), 1953 g., TsGANTD Samara (Central State Archive of Scientific and Technical Documentation in Samara), F. R-478, Op. 3–1, D. 504, LL. 3–4.

26. V. S. Fiofanov, "Ergonomika v lesnoi promyshlennosti," *Lesnaya promyshlennost'*, no. 3 (March 1990): 4; Otchet "Izuchenie osnovnykh napravleniy v razvitii tekhniki i tekhnologii lesorazrabotok," TsGANTD Samara, F. R-478, Op. 3–1, D. 505, LL. 2, 31.

27. Elena Kochetkova, "Modernization of Soviet Pulp and Paper Industry and Technology Transfer in 1953–1964: The Case of Enso/Svetogorsk," *Laboratorium: Russian Review of Social Research* 5, no. 3 (September 2013): 13–42.

28. Brenton M. Barr and Kathleen Braden, *The Disappearing Russian Forest: A Dilemma in Soviet Resource Management* (Washington, DC: Rowman and Littlefield Inc., 1988), 252.

29. "Japan-USSR Agreement for the Development of Soviet Forestry," *International Legal Materials* 8, no. 1 (1969), 48–55.

30. Nikolay Goncharenko, *Tekhnologicheskie zapasy v lesozagotovitel'noi promyshlennosti. Obzornaya informatsiya* (Moscow: Lesnaya promyshlennost', 1979), 7; Kathleen Braden, "The Role of Imported Technology in the Export Potential of Soviet Forest Products," *Association of American Geographers Project on Soviet Natural Resources in the World Economy*, no. 16 (Washington, DC, 1979), 66; Robert Jensen, Theodore Shabad, and Arthur Wright, *Soviet Natural Resources in the World Economy* (Chicago: University of Chicago Press, 1983), 463.

31. Desyatiletniy uchet lesnykh kul'tur v 1963 godu, LOGAV (Leningrad Regional State Archive in Vyborg), F. R-1659, Op. 2, D. 104, L. 424.

32. Voitto Pölkki, *Venäjän puu: puutulvasta puutulleihin* (Hämeenlinna: Kariston Kirjapaino, 2008), 208.

33. Peter Blandon, *Soviet Forest Industries* (New York: Routledge, 1983), 176–177.

34. Izuchenie spetsodezhdy, obuvi i zashchitnykh golovnykh uborov, primenyaemykh na lesozagotovkakh v SSSR i za rubezhom na predmet izyskaniya bolee ratsional'nykh gigienicheskikh obraztsov, TsGANTD Samara, F. R-478, Op. 3–1, D. 811, L. 73.

35. Nikolay Timofeev, *Osvoenie lesnykh bogatstv* (Moscow: Lesnaya promyshlennost', 1979), 5.

36. Doklad po lesnoi promyshlennosti po ekonomicheskim rayonam RSFSR, TsGANTD SPb, F. 67, Op. 1–6, D. 107, L. 8.

37. Glavnoe upravlenie lesnogo khozyaistva i okhrany lesa pri SM RSFSR; Ministerstvo lesnogo khozyaystva RSFSR. Annotatsiya, GARF (State Archive of the Russian Federation), F. A510, Op. 1, L. 2.

38. Sostoyanie tekhniki i tekhnologii proizvodstva i predvaritel'nye osnovnye napravleniya ikh razvitiya v 1959–1975 gg. po bumazhnoi i derevoobrabatyvayushchei promyshlennosti RSFSR. Goskomitet nauki i tekhniki, 1960 g., GARF, F. 408, Op. 1, D. 217, LL. 1a, 4.

39. Anatoliy Averbukh and Kseniya Bogushevskaya, *Chto delaet khimiya iz drevesiny* (Moscow: Lesnaya promyshlennost', 1970), 13.

40. A. P. Vikulov, "O tekhnicheskom progresse v tsellyulozno-bumazhnoi promyshlennosti," in *V bor'be za tekhnicheskiy progress* (Leningrad: Lenizdat, 1965), 95.

41. Elena Antropova, Aleksandra Balachenkova, Mikhail Busygin, and Vladimir Chuiko, *Istoriya tsellyulozno-bumazhnoi promyshlennosti Rossii* (Arkhangelsk: Pravda Severa, 2008), 6.

42. N. A. Lur'e, "Lesnye materialy na mirovom rynke," *Lesnaya promyshlennost'*, no. 10 (October 1965): 31.

43. Plan postavki delovoi drevesiny na eksport v Finlyandiyu i perepiski s ob'edineniem "Eksportles" po voprosam postavki drevesiny, NARK, F. 3432, Op. 1, D. 223, L. 12.

44. M. V. Malyshev, "Izuchenie utopa i kharaktera prochikh poter' lesomaterialov na lesosplave," 1946 god, TsGANTD SPb, F. 218, Op. 1–1, D. 1179, LL. 3, 17.

45. Doklad o sostoyanii tekhnicheskogo urovnya proizvodstva na lesosplave SSSR v sopostavlenii s peredovymi dostizheniyami mirovoi praktiki, 1977 god, TsGANTD SPb, F. 218, Op. 2–2, D. 107, L. 3–4.

46. A. S. Kuz'michev, S. G. Sinitsyn, and G. G. Kuznetsov, *Osnovnye tendentsii i napravleniya razvitiya lesopol'zovaniya, proizvodstva i potrebleniya drevesiny. Obzornaya informatsiya VNIITslesresursov Goskomlesa SSSR* (Moscow: VNIITslesresursov, 1991), 12.

47. Oleg Krassov, *Pravo lesopol'zovaniya v SSSR* (Moscow: Nauka, 1990), 7.

48. Materialy k zasedaniyu NTS po voprosu "Iskhodnye dannye dlya proektirovaniya lesopromyshlennykh kompleksov" ot 7 fevralya 1968 goda, RGAE (Russian State Archive of Economics), F. 73, Op.1, D. 2022, L. 7.

49. K dokladnoi zapiske rukovodstva LENNIILKh o sostoyanii i dal'neishem razvitii lesnogo khozyaistva, TsGANTD SPb, F. 310, D. 876, Op. 1–2, L. 1–2.

50. V. M. Shiryaev, "Kto sokhranit i umnozhit lesa?," *Lesnaya promyshlennost'*, no. 9 (September 1990): 8.

51. Nikolay Sinyaev, *Lesnoi kompleks Karelii: etapy perestroiki* (Petrozavodsk: Kareliya, 1990), 11, 60.

52. A. L. Tsernes, "O zadachakh i funktsiyakh promyshlennykh ob'edineniy," *Derevoobrabatyvayushchaya promyshlennost'*, no. 8 (August 1972): 1.

53. Ilya Shegel'man and Oleg Kulagin, "Osnovnye problemy razvitiya lesnoi promyshlennosti SSSR v period 1970-kh–nachala 1980-kh gg.," in *Istoricheskie, filosofskie, politicheskie i yuridicheskie nauki, kul'turologiya i iskusstvovedenie. Voprosy teorii i praktiki*, no. 6 (Tambov: Gramota, 2011), 224.

54. Proekt osnovnykh napravleniy razvitiya lesnoi, derevoobrabatyvayushchei i tsellyulozno-bumazhnoi promyshlennosti na 76–90 gg. 14 iyunya 1973 g., RGAE, F. 4372, Op. 66, D. 5850, L. 20.

55. Opis' 2, GARF, F. A510, Op. 2.

56. Doklad inspektsii lesnogo khozyaistva i okhrana lesa Leningradskoi oblasti v rabotakh na 1963 g., GARF, F. A510, Op. 2, D. 883, L. 59.

57. Serdyukov, *Prirodnye bogatstva Karelii*, 11.

58. Doklad inspektsii lesnogo khozyaistva i okhrana lesa Leningradskoi oblasti v rabotakh na 1963 g., GARF, F. A510, Op. 2, D. 883, L. 29 [emphasis added].

CHAPTER 2

1. Nikolay Timofeev, *Osvoenie lesnykh bogatstv* (Moscow: Lesnaya promyshlennost', 1979), 5.

2. E. S. Romodanovskaya, *Yakuty i ikh strana* (Moscow: Kn. magazin torgovogo doma, 1901), 11.

3. F. V. Kolosov, "Lesa na sluzhbe narodnomu khozyaistvu," *Lesnaya promyshlennost'*, no. 2 (February 1968): 28.

4. Postanovlenie regional'nogo soveshchaniya po razvitiyu proizvoditel'nykh sil Amurskoi oblasti, 26–29 marta 1962 goda, RGAE (Russian State Archive of Economics), F. 9480, Op. 7, D. 350, L. 221 [emphasis added].

5. M. Kanevskiy, "O razvitii lesnoi promyshlennosti Dal'nego Vostoka," *Lesnaya promyshlennost'*, no. 4 (April 1964): 26.

6. "Lesnye bogatstva—na sluzhbu narodnomu khozyaistvu," *Lesnaya promyshlennost'*, no. 1 (January 1960): 31.

7. G. V. Krylov and Yu. I. Khol'kin, "Voprosy khimicheskoi pererabotki listvennoi drevesiny v Sibiri," *Lesnaya promyshlennost'*, no. 10 (October 1964): 20.

8. Elena Kochetkova, "Between Water Pollution and Protection in the Soviet Union, Mid-1950s–60s: Lake Baikal and River Vuoksi," *Water History* 10, no. 2–3 (February 2018): 223–241.

9. R. Momchuk, "Organizovat' proizvodstvo khvoino-vitaminnoi muki na Dal'nem Vostoke," *Lesnaya promyshlennost'*, no. 7 (July 1961): 16.

10. A. N. Pryazhnikov and G. N. Lavrovskiy, "Perspektivy khozyaistvennogo osvoeniya kedrovykh lesov Gornogo Altaya," *Lesnaya promyshlennost'*, no. 6 (June 1962): 28.

11. Nikolay Timofeev, "Lesopol'zovanie: puti sovershenstvovaniya," *Lesnaya promyshlennost'*, no. 7 (July 1980): 1.

NOTES

12. *Lesnoe khozyaistvo SSSR za 50 let. Gosudarstvennyi komitet lesnogo khozyaistva Soveta Ministrov SSSR* (Moscow: Lesnaya promyshlennost', 1967), 312.

13. N. P. Vtorushin, "Respublika zelenogo zolota na pod'eme," *Lesnaya promyshlennost'*, no. 6 (June 1960): 11.

14. P. V. Vasil'ev and N. V. Nevzorov, "Perspektivy osvoeniya lesov Vostochnoi Sibiri," *Lesnaya promyshlennost'*, no. 9 (September 1971): 13.

15. G. Sychevskiy, "Lesnuyu otrasl' novykh rayonov—na sovremennuyu osnovu," *Lesnaya promyshlennost'*, no. 9 (September 1971): 14 [emphases added].

16. Timofeev, "Lesol'zovanie," 1.

17. A. Knopov, "O kombinirovanii lesozagotovitel'nykh, derevoobrabatyvayushchikh i tsellyulozno-bumazhnykh predpriyatiy," *Lesnaya promyshlennost'*, no. 6 (June 1957): 1.

18. V. G. Dostal', "Razvitie lesnoi promyshlennosti v Zapadnosibirskom komplekse," *Lesnaya promyshlennost'*, no. 1 (January 1967): 29.

19. N. I. Orlov, *Ispol'zovanie rek Irkutskoi oblasti dlya lesosplava*, 231, accessed November 23, 2021, http://irkipedia.ru/content/konferenciya_po_izucheniyu_proizvoditelnyh_sil_irkutskoy_oblasti.

20. Il'ya Shegel'man, *Lesnye transformatsii (XV–XXI vv.)* (Petrozavodsk: PetrGU, 2008), 95.

21. D. I. Bezmyatezhnykh, *Voprosy osvoeniya lesov severnykh rayonov Irkutskoi oblasti*, accessed February 4, 2023, http://irkipedia.ru/content/konferenciya_po_izucheniyu_proizvoditelnyh_sil_irkutskoy_oblasti.

22. I. Sandik, "Eto nas bespokoit," *Les—stroikam*, no. 1 (January 1979): 1.

23. "Obrashchenie komsomol'tsev i molodezhi Svetogorskogo TsBK k molodezhi, rabotayushchei na lesozagotovitel'nykh predpriyatiyakh Minlesproma," *Svetogorskiy rabochiy*, no. 51 (August 1956): 1.

24. V. P. Tatarinov, *Lesozagotovki: Sostoyanie i problemy povysheniya effektivnosti* (Moscow: Lesnaya promyshlennost', 1979), 71.

25. Otchet "Izuchenie i obobshchenie otechestvennogo i zarubezhnogo opyta proektirovaniya i stroitel'stva lesokhozyaistvennykh

avtomobil'nykh dorog v zonakh s intensivnym vedeniem lesnogo khozyaistva," 1967 god, TsGANTD Samara (Central State Archive of Scientific and Technical Documentation in Samara), F. R-216, Op. 2–1, D. 37, LL. 4, 7, 21, 24.

26. M. Finitskiy and B. Bozhenkov, Otchet "Obobshchenie opyta stroitel'stva lesokhozyaistvennykh dorog i ob'ektov khozyaistvennogo stroitel'stva v leskhozakh," 1962 g., TsGANTD Samara, F. R-190, Op. 2–1, D. 17, L. 69.

27. Nikolay Timofeev, "Industriya lesa v novoi pyatiletke," *Lesnaya promyshlennost'*, no 4 (April 1971): 2.

28. M. I. Brik, "Dorogi—problema nomer odin," *Lesnaya promyshlennost'*, no. 8 (August 1979): 4.

29. Ilya Shegel'man and Oleg Kulagin, "O vklade lesnogo sektora v ekonomiku SSSR v period Velikoi Otechestvennoi voiny (1941–1945 gg.)," *Nauchnyi zhurnal KubGAU* 78 (April 2012): 1–10.

30. F. A. Pavlov and A. S. Vishnyakov, "Lesovoznym dorogam—spetsializirovannuyu tekhniku," *Lesnaya promyshlennost'*, no. 8 (August 1981): 19.

31. A. Dorofeev and N. Chelyushkin, "Perspektivy razvitiya lesovoznykh avtodorog," *Lesnaya promyshlennost'*, no. 1 (January 1971): 19.

32. M. I. Kishinskiy, "Mekhanizatsiya ukhoda za avtomobil'nymi lesovoznymi dorogami," 1955 god, TsGANTD Samara, F. R-478, Op. 3–1, D. 663, L. 19.

33. I. P. Ermolin, "Na stykakh lesozagotovok i lesnogo khozyaistva," *Lesnaya promyshlennost'*, no. 8 (August 1975): 17–18.

34. M. I. Brik, "Stroit' dorogi kruglyi god," *Lesnaya promyshlennost'*, no. 11 (November 1974): 1.

35. N. A. Medvedev and A. S. Ditkovskiy, "Ispol'zovat' rezervy, vypolnit' plan," *Lesnaya promyshlennost'*, no. 2 (February 1962): 1.

36. Knopov, "O kombinirovanii lesozagotovitel'nykh, derevoobrabatyvayushchikh i tsellyulozno-bumazhnykh predpriyatiy."

37. Loren Graham, *The Ghost of the Executed Engineer: Technology and the Fall of the Soviet Union* (Cambridge, MA: Harvard University Press, 1996), 154.

38. N. A. Medvedev, "Les i ekonomika, god 1970," *Lesnaya promyshlennost'*, no. 1 (January 1970): 1.

39. B. Tikhomirov, "Nauchnaya mysl' na sluzhbe mekhanizatsii lesozagotovok v Sibiri," *Lesnaya promyshlennost'*, no. 11 (November 1948): 22.

40. *Direktivy XXIII s'ezda KPSS po pyatiletnemu planu razvitiya narodnogo khozyaistva SSSR na 1966–1970 gody* (Moscow: Izdatel'stvo politicheskoi literatury, 1966), 80.

41. L. K. Panov, *Nauchnyi otchet po teme "Izuchit' sovremennoe sostoyanie gorodov SSSR, praktiku razrabotki i realizatsii proektov rayonnoi planirovki i zastroiki gorodov."* Svodnyi doklad za 1976–1980 gg. T. 3 (Ust'-Ilimsk, 1980), 5.

42. Panov, *Nauchnyi otchet*, 9.

43. Zoya Uchastkina, *Razvitie bumazhnogo proizvodstva v Rossii* (Moscow: Lesnaya promyshlennost', 1972), 19.

44. P. P. Dorozhkin, "Uluchshit' rabotu lesozagotovitel'nykh predpriyatiy Dal'nego Vostoka," *Lesnaya promyshlennost'*, no. 5 (May 1962): 10.

45. Aleksandr Yakunin, Vladimir Shlykov, and Vladimir Grachev, *Lesnaya industriya Dal'nego Vostoka* (Moscow: Lesnaya promyshlennost', 1987), 18.

46. I. M. Sen'kin, "Lesopil'naya i derevoobrabatyvayushchaya promyshlennost' Vostochnoi Sibiri," *Derevoobrabatyvayushchaya promyshlennost'*, no. 11 (November 1967): 24.

47. N. V. Nevzorov, "Lesozagotovki v SSSR i ikh razmeshchenie," *Lesnaya promyshlennost'*, no. 10 (October 1957): 5; *Lesnoe khozyaistvo SSSR za 50 let* (Moscow: Izdatel'stvo "Lesnaya promyshlennost'," 1967).

48. V. Mozhin, "Ratsional'noe razmeshchenie proizvoditel'nykh sil i sovershenstvovanie territorial'nykh proportsiy," *Planovoe khozyaistvo*, no. 4b (1983): 4–5.

49. Mark Bandman, *Territorial'no-proizvodstvennye kompleksy: teoriya i praktika predplanovykh issledovaniy* (Novosibirsk: Nauka, 1980), 12.

50. A. A. Filatov, "Kompleksnye lesokombinaty—mnogolesnym rayonam," *Lesnaya promyshlennost'*, no. 12 (December 1980): 15.

51. N. G. Bagaev, "Lesnye kompleksy—predpriyatiya budushchego," *Lesnaya promyshlennost'*, no. 4 (April 1981): 9.

52. "V gosudarstvennom nauchno-tekhnicheskom komitete Soveta ministrov RSFSR," *Derevoobrabatyvayushchaya promyshlennost'*, no. 5 (May 1960): 27.

53. A. V. Popov, "Proizvodstvo stalo bezotkhodnym," *Lesnaya promyshlennost'*, no. 5 (May 1981): 8.

54. Elena Kochetkova and Aleksey Popov, "Socialist Construction for Siberia: Comecon Cooperation and the Making of Ust'-Ilimsk Forest Industrial Complex in the USSR, 1970s and 1980s," *Journal of Contemporary History* 57, no. 2 (2022): 479–498.

55. Viktor Kalinkin, *Gigant na Angare* (Moscow: Lesnaya promyshlennost', 1984), 25–27.

56. Irina Koryukhina, Tat'yana Timofeeva, Tat'yana Grebenshchikova, Irina Abdulova, Vera Kuklina, and Mikhail Rozhanskiy, *Gorod posle kombinata: Sotsial'no-ekonomicheskie strategii zhitelei goroda Baikal'ska* (Irkutsk: TsNSI, 2012), 101. For more on the development of Siberia, see, for example, Al'bina Timoshenko, "Industrial'noe osvoenie Sibiri vo vtoroi polovine XX v.: frontirnoe izmerenie," *Ural'skiy istoricheskiy vestnik* 61, no. 4 (2018): 104–111.

57. *Razvitie proizvoditel'nykh sil Sibiri. Lesnoe khozyaistvo i lesnaya promyshlennost'* (Moscow: Izdatel'stvo Akademii nauk, 1960), 229.

58. Kanevskiy, "O razvitii lesnoi promyshlennosti Dal'nego Vostoka," 26.

59. *Obzory po informatsionnomu obespecheniyu tselevykh kompleksnykh nauchno-tekhnicheskikh program i program po resheniyu vazhneishikh nauchno-tekhnicheskikh problem. Obzornaya informatsiya* (Moscow: TsBNTI, 1991), 8.

60. A. Chilimov and A. Tsekhmistrenko, "Intensifitsirovat' lesokhozyaistvennoe proizvodstvo," *Lesnaya promyshlennost'*, no. 2 (February 1971): 16.

61. Materialy k zasedaniyu NTS po voprosu "Iskhodnye dannye dlya proektirovaniya lesopromyshlennykh kompleksov" ot 7 fevralya 1968 g. Ministerstvo SSSR. Nauchno-tekhnicheskiy sovet, RGAE, F. 73, Op. 1, D. 2022, L. 8.

62. "Bol'shie zadachi lesnoi industrii," *Derevoobrabatyvayushchaya promyshlennost'*, no. 10 (October 1966): 1.

63. I. N. Voevoda and V. I. Klevtsov, "Sovershenstvovanie sistemy rubok v lesakh Zapadnoi Sibiri," *Lesnaya promyshlennost'*, no. 5 (May 1970): 18.

64. For more on industrial allocation in Siberia and its consequences, see Fiona Hill and Clifford G. Gaddy, *The Siberian Curse: How Communist Planners Left Russia Out in the Cold* (Washington, DC: Brookings Institution Press, 2003), 240.

65. Yakunin, Shlykov, and Grachev, *Lesnaya industriya Dal'nego Vostoka*, 55.

66. V. T. Gorbachev, "Kakim dolzhen byt' lesnoi poselok," *Lesnaya promyshlennost'*, no. 4 (April 1963), 21. For more on the destiny of monotowns after the closure of enterprises, see Irina Koryukhina et al., *Gorod posle kombinata*, 145.

67. Svodnoe zaklyuchenie, noyabr' 1983 goda, RGAE, F. 4372, Op. 67, D. 6175, L. 123.

68. A. G. Zheludko, "O ratsional'nom ispol'zovanii lesnykh resursov," *Lesnaya promyshlennost'*, no. 1 (January 1960): 13.

69. Prikazy po tsellyulozno-bumazhnoi promyshlennosti i Minbumpromu, RGAE, F. 4273, Op. 66, D. 7307, L. 76.

70. For more on earlier Soviet legislation and conservation, see Brian Bonhomme, *Forests, Peasants, and Revolutionaries: Forest Conservation and Organization in Soviet Russia, 1917–1929* (New York: Columbia University Press, 2005).

71. Otchet "Izuchenie i obobshchenie otechestvennogo i zarubezhnogo opyta proektirovaniya i stroitel'stva kompleksnykh lesnykh predpriyatiy," TsGANTD Samara, F. R-216, Op. 2–1, D. 36, L. 81.

72. P. Polyakov, "Za kompleksnoe osvoenie lesov Khabarovskogo kraya," *Lesnaya promyshlennost'*, no. 3 (March 1963): 29.

73. V. T. Nikolaenko, "Zashchitnoe lesorazvedenie i okhrana okruzhayushchei sredy," *Obzory po informatsionnomu obespecheniyu tselevykh kompleksnykh nauchno-tekhnicheskikh program* 2, no. 26 (1986): 81.

74. Sten Nilsson and Anatoly Shvidenko, "The Russian Forest Sector: A Position Paper for the World Commission on Forests and

Sustainable Development," in *United Nations Publication* (Laxenburg, 1997), 9.

75. Pertti Hari, Kari Heliövaara, and Liisa Kulmala, *Physical and Physiological Forest Ecology* (New York: Springer, 2012), 473.

76. Spravka o rezul'tatakh proverki postanovleniya VS KASSR ot 18.06.1961 goda ob okhrane prirody, NARK (National Archive of the Republic of Karelia), F. 3432, Op. 1, D. 242, L. 6.

77. Nikolay Sinyaev, *Lesnoi kompleks Karelii: etapy perestroiki* (Petrozavodsk: Kareliya, 1990), 9.

78. Brenton M. Barr and Kathleen Braden, *The Disappearing Russian Forest: A Dilemma in Soviet Resource Management* (Washington, DC: Rowman and Littlefield Inc., 1988), 228.

79. A. I. Buzykin, M. D. Evdokimenko, and L. S. Pshenichnikova, "Iz opyta nesploshnykh rubok v lesakh Vostochnoi Sibiri," *Aktual'nye problemy lesnogo kompleksa*, no. 12 (2005): 5.

80. Iz pis'ma pisatelya O. V. Volkova General'nomu sekretaryu TsK KPSS L. I. Brezhnevu o neratsional'noi ekspluatatsii lesnykh resursov strany, RGAE, F. 4372, Op. 67, D. 818, LL. 2–16.

81. A. V. Beigel'man, "Tsellyuloza, bumaga i karton. Obzornaya informatsiya," in *Lesosyr'evye resursy Sibiri i Dal'nego Vostoka dlya tsellyulozno-bumazhnoi promyshlennosti*, ed. A. V. Beigel'man, T. N. Antoshina, and A. M. Gasinets, no. 7 (Moscow: VNIPIEllesprom, 1982), 3.

82. Razrabotka tekhnologicheskogo protsessa oblagorazhivaniya drovyanoi drevesiny, TsGANTD Samara, F. R-478, Op. 3–1, D. 784, L. 5.

83. Philip R. Pryde, *Environmental Management in the Soviet Union* (Cambridge: Cambridge University Press, 1991), 121.

84. N. A. Moiseev, *Osnovy prognozirovaniya, ispol'zovaniya i vosproizvodstva resursov* (Moscow: Lesnaya promyshlennost', 1974), 168–185.

85. V. I. Loshakov, "Basseinu Baikala—ratsional'noe ispol'zovanie," *Lesnaya promyshlennost'*, no. 10 (October, 1988): 30.

86. I. P. Bardin, ed., *Razvitie proizvoditel'nykh sil Vostochnoi Sibiri: Trudy konferentsii v 13-ti tomakh. Lesnoe khozyaistvo i lesnaya promyshlennost'* (Moscow: Izd-vo AN SSSR, 1960), 197.

87. N. S. Savchenko, "Dela i nuzhdy dal'nevostochnykh lesozagotovitelei," *Lesnaya promyshlennost'*, no. 2 (February 1962): 4.

CHAPTER 3

1. Here I build on the argument of Michael David-Fox that ideology is not opposite to rationality but instead part of it. Michael David-Fox, *Crossing Borders: Modernity, Ideology, and Culture in Soviet Russia* (Pittsburgh: University of Pittsburgh Press, 2015), 296.

2. See Andy Bruno, *The Power of Nature: An Arctic Environmental History* (Cambridge: Cambridge University Press, 2016), 271.

3. *Syr'evye resursy bumazhnoi promyshlennosti* (Moscow: Goslestekhizdat, 1932).

4. M. I. Brik, "Okhrana okruzhayushchei sredy i ratsional'noe ispol'zovanie prirodnykh resursov v lesnoi, tsellyulozno-bumazhnoi i derevoobrabatyvayushchei promyshlennosti," *Lesopol'zovanie i puti snizheniya ego vliyaniya na okruzhayushchuyu sredu* (Moscow: VNIIPIE, 1985), 2:1–2 [emphasis added].

5. "Zadachi lesopil'no-derevoobrabatyvayushchei promyshlennosti v 1958 g.," *Lesnaya promyshlennost'*, no. 6 (June 1958): 2.

6. Postanovlenie TsK KPSS i SM SSSR "Ob uporyadochenii ispol'zovaniya i usileniya okhrany prirodnykh resursov," 1970 god, RGAE (Russian State Archive of Economics), F. 4372, Op. 66, D. 5424, L. 6.

7. Materialy k zasedaniyu NTS po voprosu "Iskhodnye dannye dlya proektirovaniya lesopromyshlennykh kompleksov" ot 7 fevralya 1968 g. Ministerstvo SSSR. Nauchno-tekhnicheskiy sovet, RGAE, F. 73, Op. 1, D. 2022, L. 6.

8. "Puti kompleksnogo osvoeniya lesnykh bogatstv," *Lesnaya promyshlennost'*, no. 12 (December 1961): 2.

9. N. G. Sud'ev, "Lesnym resursam—kompleksnoe ispol'zovanie," *Lesnaya promyshlennost'*, no. 5 (May 1973): 2 [emphasis added].

10. V. F. Zaretskiy, "Vsemerno ekonomit' toplivo i energiyu!," *Derevoobrabatyvayushchaya promyshlennost'*, no. 9 (September 1979): 3.

11. "Strogo soblyudat' normy raskhoda syr'ya i materialov," *Derevoobrabatyvayushchaya promyshlennost'*, no. 12 (December 1954): 1; "Lesnye bogatstva Vostochnoi Sibiri na sluzhbu narodnomu khozyaistvu," *Lesnaya promyshlennost'*, no. 7 (July 1958): 27.

12. "Vysokaya lichnaya otvetstvennost' i initsiativa rukovoditelya," *Les–stroikam*, no. 33 (August 1979): 1.

13. D. I. Ivanov, "Razvitie tsellyulozno-bumazhnoi promyshlennosti v poslevoennyi period," *Trudy Leningradskogo tekhnologicheskogo instituta tsellyulozno-bumazhnoi promyshlennosti*, no. 15 (1965): 41.

14. *Perspektivnaya otsenka zapasov lesnykh resursov i obespechennosti imi narodnogo khozyaistva SSSR* (Moscow, 1979), 1. See also Otchet "Izuchenie otechestvennogo i zarubezhnogo opyta proektirovaniya i stroitel'stva kompleksnykh lesnykh predpriyatiy," TsGANTD (Central State Archive of Scientific and Technical Documentation), F. R-216, Op. 2–1, D. 36, L. 81.

15. Pis'mo predsedatelyu GNTK SM SSSR K. I. Petukhovu ot direktora Enso Olontseva, LOGAV (Leningrad Regional State Archive in Vyborg), F. R-180, Op. 5, D. 180, L. 16.

16. *Perspektivnaya otsenka zapasov lesnykh resursov i obespechennosti imi narodnogo khozyaistva SSSR* (Moscow: Sovet po izucheniyu proizvoditel'nykh sil, 1979), 6.

17. V. D. Solomonov, "Effektivnee ispol'zovat' drevesnoe syr'e," *Derevoobrabatyvayushchaya promyshlennost'*, no. 1 (January 1985): 1.

18. For more on theory on waste, see Zsuzsa Gille, "Actor Networks, Modes of Production, and Waste Regimes: Reassembling the Macro-Social," *Environment and Planning: Economy and Space*, no. 42 (2010): 1049–1064.

19. Karl Marx, *Capital: A Critique of Political Economy* (Moscow: Progress Publishers, 2015), 1:27.

20. E. L. Nordshtrem and A. A. Lizunov, "Ob ispol'zovanii otkhodov lesozagotovok," *Lesnaya promyshlennost'*, no. 12 (December 1949): 21.

21. S. Kochubei and V. Kislyi, "Gde zabota ob ispol'zovanii otkhodov," *Lesnaya promyshlennost'*, no. 4 (April 1964): 24.

22. "Bol'shoi khimii—udarnye tempy!," *Master lesa*, no. 10 (October 1963): 1.

23. Jeronim Perović, *Cold War Energy: A Transnational History of Soviet Oil and Gas* (London: Palgrave Macmillan, 2017), 454.

24. P. Pyudik et al., Otchet "Sovershenstvovanie tekhnologii proizvodstva drevesnykh plastikov," 1965 god, TsGANTD, F. 84, Op. 2–1, D. 565, L. 3.

25. A. G. Rakin et al., Tekhnicheskiy otchet po teme "Razrabotka tekhnologii drevesnogo plastika iz otkhodov drevesiny bez svyazuyushchikh," TsGANTD, F. 84, Op. 2–1, D. 566, L. 2a.

26. A. S. Nikiforov, "Proizvodstvu drevesnykh plit—preimushchestvennoe razvitie," *Derevoobrabatyvayushchaya promyshlennost'*, no. 7 (July 1962): 1.

27. B. A. Vasyl'ev, "Mebel' komponuet pokupatel'," *Derevoobrabatyvayushchaya promyshlennost'*, no. 6 (June 1987): 25–27.

28. A. F. Smolyakov, A. B. Dobrov, A. K. Leont'ev, and N. V. Poshenev, "Gazoobraznoe toplivo iz drevesnykh otkhodov," *Lesnaya promyshlennost'*, no. 12 (December 1987): 16.

29. A. S. Sinnikov, A. A. Shamin, A. L. Parshevnikov, and L. V. Menshchikova, "Ob ispol'zovanii drevesnoi kory," *Derevoobrabatyvayushchaya promyshlennost'*, no. 11 (November 1976): 12.

30. "Suveniry iz otkhodov mebel'nogo i derevoobrabatyvayushchego proizvodst," *Derevoobrabatyvayushchaya promyshlennost'*, no. 2 (February 1972): 33.

31. Mikhail Serdyukov, *Prirodnye bogatstva Karelii na sluzhbu narodu* (Petrozavodsk: Kareliya, 1959), 14.

32. E. I. Lopukhov, "Voprosy razvitiya lesnoi promyshlennosti v Bratskom rayone Irkutskoi oblasti," *Lesnaya promyshlennost'*, no. 12 (December 1956): 5.

33. Zasedanie sektsii nauchnogo soveta komiteta po koordinatsii NIR po probleme "Kompleksnoe ispol'zovanie i vosproizvodstvo lesnykh resursov i nedrevesnogo rastitel'nogo syr'ya," 20.02.1962, RGAE, F. 9480, Op. 7, D. 533, LL. 34–35 [emphasis added].

34. *Sobirat' makulaturu—sokhranyat' prirodnye bogatstva*, 1982, accessed November 23, 2021, https://www.net-film.ru/found-page-1/?search=q %d1%81%d0%b1%d0%be%d1%80%2B%d0%bc%d0%b0%d0%ba %d1%83%d0%bb%d0%b0%d1%82%d1%83%d1%80%d1%8b.

35. Tat'yana Kostitsyna, *Nam s toboi povezlo, Ust'-Ilimsk!* (Syktyvkar: OOO "Infotsentr," 2020), 5.

36. G. M. Benenson, "Problemy ispol'zovaniya otkhodov drevesiny na soveshchanii v Akademii nauk SSSR," *Lesnaya promyshlennost'*, no. 8 (August 1955): 31.

37. Stenograficheskiy otchet ekonomicheskoi sektsii uchenogo soveta instituta o resursakh makulatury i uluchshenii ee ispol'zovaniya v bumazhnoi promyshlennosti, 1956 god, TsGANTD SPb (Central State Archive of Scientific and Technical Documentation in Saint Petersburg), F. 303, Op. 12, D. 284, LL. 2, 5. See also L. V. Ross, "Doklad na chitatel'skoi konferentsii zhurnala 'Lesnaya promyshlennost'," 1957 god, TsGANTD SPb, F. 342, Op. 1–1, D. 229, L. 116.

38. Stenograficheskiy otchet ekonomicheskoi sektsii uchenogo soveta instituta o resursakh makulatury, LL. 7, 19, 35 [emphasis added].

39. Stenograficheskiy otchet ekonomicheskoi sektsii uchenogo soveta instituta o resursakh makulatury, LL. 7, 19, 35.

40. Konferentsiya po voprosu "Sovremennaya tekhnologiya i oborudovanie po pererabotke makulatury i ispol'zovanie ee v proizvodstve kartona i bumagi," 20.10.1971, TsGANTD SPb, F. 367, Op. 1–2, D. 170, L. 47.

41. Protokol obshchego partiynogo sobraniya partiynoi organizatsii Svetogorskogo TsBK ot 20.09.1956, TsGAIPD (Central State Archive of Historical and Political Documentation), F. R-1542, Op. 4, D. 2, L. 2.

42. Konferentsiya po voprosu "Sovremennaya tekhnologiya i oborudovanie po pererabotke makulatury i ispol'zovanie ee v proizvodstve kartona i bumagi," 21.10.1971, TsGANTD SPb, F. 367, Op. 1–2, D. 171, L. 24.

43. Stenograficheskiy otchet ekonomicheskoi sektsii uchenogo soveta instituta o resursakh makulatury, LL. 57–58.

44. Konferentsiya po voprosu, L. 32.

45. D. Kovaleva, "Povyshat' effektivnost'," *Vyborgskiy kommunist* (August 19, 1982): 2.

46. Elena Antropova, Aleksandra Balachenkova, Mikhail Busygin, and Vladimir Chuiko, *Istoriya tsellyulozno-bumazhnoi promyshlennosti Rossii* (Arkhangelsk: Pravda Severa, 2008), 15.

47. Konferentsiya po voprosam "Sovremennaya tekhnologiya i oborudovanie po pererabotke makulatury," 20.10.1971, TsGANTD, F. 367, Op. 1–2, D. 170, L. 47.

48. M. V. Malyshev, "Izuchenie utopa i kharaktera prochikh poter' lesomaterialov na lesosplave," 1946 god, TsGANTD SPb, F. 218, Op. 1–1, D. 1179, L. 18.

49. Nikolay Timofeev, *Lesnaya industriya SSSR* (Moscow: Lesnaya promyshlennost', 1980), 3.

50. Dmitriy Sokolovskiy, *Sbor i pererabotka makulatury na bumagu i karton* (Moscow: Goslesbumizdat, 1957), 6. See also "Zadachi lesopil'no-derevoobrabatyvayushchei promyshlennosti v 1958 godu," *Lesnaya promyshlennost'*, no. 6 (June 1958): 2.

51. R. Tomchuk, "Kompleksno ispol'zovat' lesosyr'evye resursy," *Lesnaya promyshlennost'*, no. 11 (November 1970): 27.

52. Proekt osnovnykh napravleniy razvitiya lesnoi, derevoobrabatyvayushchei i tsellyulozno-bumazhnoi promyshlennosti na 1975–1990 gg., 14.06.1973, RGAE, F. 4372, Op. 66, D. 5850, L. 5.

53. A. V. Popov and V. M. Khmilevskiy, "Kompleksnoe ispol'zovanie drevesiny na Beregometskom lesokombinate," *Derevoobrabatyvayushchaya promyshlennost'*, no. 7 (July 1980): 20.

54. B. M. Kudryavtsev, "Otkhodov proizvodstva—net!," *Derevoobrabatyvayushchaya promyshlennost'*, no. 10 (October 1977): 17.

55. E. N. Bykov, "Kak ispol'zovat' drevesinu such'ev," *Lesnaya promyshlennost'*, no. 9 (September 1982): 19.

56. V. P. Tseplyaev, *Vazhneishie polozheniya i printsipy sovetskogo lesopol'zovaniya* (Moscow: TsBNTI Gosleskhoza SSSR, 1979), 33–34 [emphasis added].

57. Porucheniya Soveta ministrov SSSR i materialy po vypolneniyu porucheniy po voprosam lesnogo kompleksa, 1984 god, RGAE, F. 9480, Op. 13, D. 1581, L. 47.

58. Aleksandr Yakunin, Vladimir Shlykov, and Vladimir Grachev, *Lesnaya industriya Dal'nego Vostoka* (Moscow: Lesnaya promyshlennost', 1987), 37.

59. Sanitarnyi obzor po Gatchinskomu leskhozu za 1952 god, LOGAV, F. R-1659, Op. 2, D. 12, L. 8.

60. Yu. D. Abaturov, *Lesnye resursy SSSR: Sostoyanie i okhrana* (Moscow: VNTITsentr, 1985), 50.

61. Nikolay Sinyaev, *Lesnoi kompleks Karelii: etapy perestroiki* (Petrozavodsk: Kareliya, 1990), 64.

62. Porucheniya Soveta ministrov SSSR, LL. 1–3.

63. K proektu postanovleniya SM SSSR "O merakh po dal'neishemu razvitiyu gosudarstvennoi torgovli," 1975 god, RGAE, F. 4372, Op. 66, D. 7306, L. 48.

64. Zasedanie sektsii nauchnogo soveta komiteta po koordinatsii NIR po probleme "Kompleksnoe ispol'zovanie i vosproizvodstvo lesnykh resursov i nedrevesnogo rastitel'nogo syr'ya," 20.02.1962, RGAE, F. 9480, Op. 7, D. 533, L. 189.

65. Pis'mo nachal'nika otdela lesnoi i tsellyuozno-bumazhnoi promyshlennosti V. P. Tatarinova, sentyabr' 1975 goda, RGAE, F. 4372, Op. 66, D. 7307, L. 60.

66. *Okhrana okruzhayushchei sredy i ratsional'noe ispol'zovanie prirodnykh resursov* (Moscow: VNTITsentr, 1986), 2.

67. See Astrid M. Kirchhof and John McNeill, eds., *Nature and the Iron Curtain: Environmental Policy and Social Movements in Communist and Capitalist Countries, 1945–1990* (Pittsburgh: University of Pittsburgh Press, 2019), 216.

68. A. I. Shcherbakov, "Po-novomu ispol'zovat', bystree vozobnovlyat' les," *Lesnaya promyshlennost'*, no. 9 (September 1964): 6.

69. Nikolay Anuchin, *Teoriya i praktika organizatsii lesnogo khozyaistva* (Moscow: Izd-vo "Lesnaya promyshlennost'," 1977).

70. Effektivnost' razmeshcheniya lesopromyshlennogo proizvodstva i otsenka lesnykh resursov. Sbornik statei, 1975 g. Sovet po izucheniyu proizvodstvennykh sil pri Gosplane SSSR, RGAE, F. 399, Op. 3, D. 1445, L. 71.

71. Vasiliy Sokolov and Grigoriy Khozin, *Promyshlennost' i okhrana okruzhayushchei sredy* (Moscow: Znanie, 1978), 22–23.

72. Stanislav Sinitsyn, *Ratsional'noe lesopol'zovanie* (Moscow: Agropromizdat, 1987), 278.

73. B. Kyucharyants, "Priroda ili okruzhayushchaya sreda?," *Les—stroikam*, no. 39 (October 1981): 2.

74. Nikolay Timofeev, *Lesnaya industriya* (Moscow: Lesnaya promyshlennost', 1980): 5.

75. *Vtorichnye material'nye resursy lesnoi i derevoobrabatyvayushchei promyshlennosti* (Moscow: Ekonomika, 1983), 6.

76. A. M. Chashchin, *Khimiya zelenogo zolota* (Moscow: Lesnaya promyshlennost', 1987), 8.

77. V. I. Parshikov, "Prirodookhrannoe zakonodatel'stvo i khozraschet," *Lesnaya promyshlennost'*, no. 6 (June 1988): 20.

78. S. Rusan, "Okhrana prirody—delo kazhdogo," *Les—stroikam*, no. 20 (May 1984): 1.

79. A. Kubenskiy, "Grazhdanskiy dolg kazhdogo cheloveka," *Vyborgskiy kommunist*, no. 23 (1983): 1.

80. Yu. A. Yagodnikov, "Problemy ekologii reshat' soobshcha," *Lesnaya promyshlennost'*, no. 9 (September 1990): 3 [emphasis added].

81. Oleg Krassov, *Pravo lesopol'zovaniya v SSSR* (Moscow: Nauka, 1990), 3.

82. G. P. Vlasov, "Chtoby ne vredit' prirode," *Lesnaya promyshlennost'*, no. 6 (June 1990): 8 [emphasis added].

83. Vlasov, "Chtoby ne vredit' prirode," 8.

84. Samuel Sprintsyn, *Ekonomika ispol'zovaniya vtorichnykh drevesnykh resursov* (Moscow: Lesnaya promyshlennost', 1990), 79.

CHAPTER 4

1. Spravka po voprosu ob utverzhdenii skorrektirovannogo proektnogo zadaniya na stroitel'stvo 1 i 2 ocheredi Astrakhanskogo TsBK, RGAE (Russian State Archive of Economics), F. 339, Op. 6, D. 3930, L. 38.

2. Zoya Uchastkina, *Razvitie bumazhnogo proizvodstva v Rossii* (Moscow: Lesnaya promyshlennost', 1972), 117.

3. V. Mudrik, "Trostnik—nash bol'shoi klad," *Master lesa*, no. 10 (October 1960): 11.

4. Ivan Popov, *Trostnikovye zarosli kak syr'evaya baza TsBK* (Moscow: Lesnaya promyshlennost', 1964), 3.

5. E. L. Akim, V. Kil'kki, and M. Ivanova, *Razrabotka tekhnologicheskogo rezhima varki i otbelki osnovnoi tsellyulozy v usloviyakh Khersonskogo tsellyuloznogo zavoda* (Leningrad, 1964), 1.

6. Soviet industry began mass production of toilet paper in the late 1960s.

7. Tekhniko-ekonomicheskoe obsledovanie rayonov del'ty Volgi (Astrakhanskoi oblasti) dlya vyyavleniya usloviy stroitel'stva tsellyulozno-bumazhnogo predpriyatiya s ispol'zovaniem v kachestve

syr'ya mestnogo kamysha, 1955 god, RGAE, F. 8513, Op. 4, D. 1468, L. 4.

8. Tekhniko-ekonomicheskoe obsledovanie rayonov del'ty Volgi, L. 4 [emphasis added].

9. Tekhniko-ekonomicheskoe obsledovanie rayonov del'ty Volgi, L. 13.

10. Direktoru TsNIIBa S. A. Puzyrevu ot direktora ukrfiliala TsNIIBa M. I. Lyskova, TsGANTD SPb (Central State Archive of Scientific and Technical Documentation in Saint Petersburg), F. 303, Op. 13, D. 247, LL. 1–2.

11. G. Kuz'mishchev, "Neskol'ko zamechaniy o kamyshe, rastushchem po Volge i Kaspiyskomu pomor'yu," *Lesnoi zhurnal* 4, no. 10 (1840): 132–139.

12. Prakticheskie predlozheniya po materialam zagranichnoi komandirovki M. M. Lyskova, V. N. Malyutina i G. M. Prokhorenko v Rumynskuyu narodnuyu respubliku po izucheniyu pererabotki trostnika na tsellyulozno-bumazhnuyu produktsiyu, TsGANTD SPb, F. 303, Op. 13, D. 802, LL. 2, 12. See also Elena Kochetkova, "Technological Inequalities and Motivation of Soviet Institutions in the Scientific-Technological Cooperation of Comecon in Europe, 1950s–80s, " *European Review of History: Revue européenne d'histoire* 28, no. 3 (2021): 355–373.

13. F. F. Derbentsev and A. F. Grishankov, *Zagotovka i khranenie trostnika v Rumynskoi narodnoi respublike* (Moscow: B.i., 1959), 1.

14. Prakticheskie predlozheniya po materialam zagranichnoi komandirovki, LL. 18, 31–32.

15. R. F. Ul'berg, *Novye zavody po vyrabotke belenoi tsellyulozy iz trostnika* (Moscow: B.i., 1964), 15.

16. Ekspertnoe zaklyuchenie po general'noi skheme osvoeniya zapasov kamysha (trostnika) v Astrakhanskoi oblasti, RGAE, F. 339, Op. 6, D. 3930, L. 32.

17. Otchet F. F. Kuteinikova "Perspektivy razvitiya tsellyulozno-bumazhnoi promyshlennosti. Utochnenie resursov i usloviy ispol'zovaniya trostnika v tsellyulozno-bumazhnoi promyshlennosti SSSR v perspektive," 1960 god, TsGANTD SPb, F. 303, Op. 2–2, D. 1422, LL. 6, 17.

18. Ekspertnoe zaklyuchenie po general'noi skheme osvoeniya zapasov kamysha (trostnika) v Astrakhanskoi oblasti, RGAE, F. 339, Op. 6, D. 3930, LL. 14, 19.

19. G. Asteryakov, "Astrakhanskiy pervenets," *Master lesa*, no. 12 (December 1963): 3.

20. Aleksey Gvishiani, *Fenomen Kosygina. Zapiski vnuka. Mnenie sovremennikov* (Ekaterinburg: Fond kul'tury "Ekaterina," 2004), 98–99. See also Elena Kochetkova, "Industry and Forests: Alternative Raw Materials in the Soviet Forestry Industry from the Mid-1950s to the 1960s," *Environment and History* 24, no. 3 (2018): 323.

21. Popov, *Trostnikovye zarosli kak syr'evaya baza TsBK*, 4 [emphasis added].

22. V. I. Mudrik, *Astrakhanskiy TsBK* (Moscow: B.i., 1960), 57.

23. L. N. Zalesskiy, "Ekspertnoe zaklyuchenie po general'noi skheme osvoeniya trostnika v Astrakhanskoi oblasti," 1962 g. RGAE, F. 339, Op. 6, D. 3930, L. 14.

24. Zaklyuchenie glavnogo upravleniya gosudarstvennoi ekspertizy Gosstroya po skorrektirovannomu proektnomu zadaniyu na stroitel'stvo 1 i 2 ocheredei Astrakhanskogo TsKK Ministerstva lesnoi, tsellyulozno-bumazhnoi i derevoobrabatyvayushchei promyshlennosti SSSR, RGAE, F. 339, Op. 6, D. 3937, LL. 246, 69.

25. Derbentsev and Grishankov, *Zagotovka i khranenie trostnika v Rumynskoi narodnoi respublike*, 13, 20.

26. Zaklyuchenie ekspertnoi podkomissii Gosudarstvennoi ekspertnoi komissii Gosplana SSSR po proetknym materialam na stroitel'stvo Astrakhanskogo i Kzyl-Ordynskogo tsellyuloznogo kombinatov, Izmail'skogo i Khersonskogo tsellyuloznogo zavodov, RGAE, F. 339, Op. 6, D. 3931, L. 69.

27. Kratkiy otchet o vypolnenii UkrNIIBom rabot po linii SEV za 1964 g., TsGANTD SPb, F. 303, Op. 13, D. 800, L. 74.

28. Gosplanu SSSR, Gosudarstvennoi ekspertnoi komissii ot nachal'nika Khersonesskogo MO upravleniya lesnogo khozyaistva i lesozagotovok, RGAE, F. 4372, Op. 65, D. 919, L. 34.

29. Vypiska iz protokola zasedaniya Byuro VSNKh SM SSSR ot 7 yanvarya 1964 goda. Ob utverzhdenii skorrektirovannogo proektnogo

zadaniya na stroitel'stvo pervoi i vtoroi ocheredi Astrakhanskogo TsKK, 1964 god, RGAE, F. 339, Op. 6, D. 3930, L. 43.

30. O sryve vypolneniya plana zagotovki trostnika dlya Astrakhanskogo TsKK i Khersonskogo tsellyuloznogo zavoda, 8 yanvarya 1964 goda, RGAE, F. 339, Op. 6, D. 3930, LL. 49–50.

31. Pravitel'stvennaya telegramma, RGAE, F. 339, Op. 6, D. 3931, L. 1.

32. Spravka po Khersonskomu tsellyuloznomu zavodu, 29.09.1964, RGAE, F. 339, Op. 6, D. 3931, L. 4.

33. Gosplanu SSSR, Gosudarstvennoi ekspertnoi komissii.

34. O sryve vypolneniya plana zagotovki trostnika.

35. Protokol sobraniya partiyno-khozyaistvennogo aktiva TsKK g. Kzyl-Orda, 24 yanvarya 1966 goda, RGAE, F. 73, Op. 1, D. 216, L. 111.

36. Khersonskiy zavod, osnovanie dlya proektirovaniya, RGAE, F. 339, Op. 6, D. 3930, L. 88.

37. Morris Bornstein, *The Soviet Economy: Continuity and Change* (Abingdon-on-Thames, UK: Routledge, 2019), 381.

38. Ob izmenenii profilya 3 ocheredi Astrakhanskogo TsKK, RGAE, F. 339, Op. 6, D. 3930, L. 170.

39. Poyasnitel'naya zapiska N. A. Guzun i T. K. Berezinskoi, RGAE, F. 339, Op. 6, D. 3930, L. 101.

40. Jenny Smith, "Agricultural Involution in the Postwar Soviet Union," *International Labour and Working-Class History*, no. 85 (Spring 2014): 59–74.

41. V. Karymsakov, pis'mo v VSNKh ot 12 avgusta 1964 goda, RGAE, F. 339, Op. 6, D. 3930, L. 224.

42. Dokladnaya V. Bogdanova, predsedatelya Gosudarstvennoi priemochnoi komissii, 7 avgusta 1967 goda, RGAE, F. 73, Op. 1, D. 2137, LL. 1, 4, 10.

43. Ob'yasnitel'naya zapiska Minbumproma k otchetu Astrakhanskogo tsellyulozno-kartonnogo kombinata po osnovnoi deyatel'nosti za 1968 god, 1969 g., RGAE, F. 442, Op. 1, D. 208, L. 7.

44. Otchet Astrakhanskogo TsKK po osnovnoi deyatel'nosti za 1979 god, 24 yanvarya 1980 g., RGAE, F. 442, Op. 1, D. 5878, LL. 94, 95.

45. M. V. Malyshev, "Izuchenie utopa i kharaktera prochikh poter' lesomaterialov na lesosplave," 1946 god, TsGANTD SPb, F. 218, Op. 1–1, D. 1179, L. 135.

46. Otchet Astrakhanskogo TsKK po osnovnoi deyatel'nosti za 1979 god, L. 113.

47. Otchet Astrakhanskogo TsKK po osnovnoi deyatel'nosti za 1980 god i ob'yasnitel'naya zapiska k otchetu, 23 yanvarya 1981 g., RGAE, F. 73, Op. 2, D. 397, L. 217.

48. I. Bodrov, "Shumyat kamyshi. Volga v opasnosti," *Lesnaya promyshlennost'*, no. 2 (February 1970): 3.

CHAPTER 5

1. "Derevoobrabatyvayushchei promyshlennosti—vysokoproizvoditel'noe oborudovanie," *Derevoobrabatyvayushchaya promyshlennost'*, no. 10 (October 1955): 2.

2. "Novaya pyatiletka sovetskoi promyshlennosti," *Derevoobrabatyvayushchaya promyshlennost'*, no. 4 (April 1966): 2.

3. David Mowery, *Paths of Innovation: Technological Change in 20th-Century America* (Cambridge: Cambridge University Press, 2000), 8.

4. Anatoliy Averbukh and Kseniya Bogushevskaya, *Chto delaet khimiya iz drevesiny* (Moscow: Lesnaya promyshlennost', 1970).

5. "Truzheniki lesa—bol'shoi khimii," *Lesnaya promyshlennost'*, no. 1 (January 1964): 1.

6. See, for example, "Likvidirovat' otstavanie lesozagotovitel'noi promyshlennosti," *Lesnaya promyshlennost'*, no. 11 (November 1953): 1.

7. Stenograficheskiy otchet ekonomicheskoi sessii Uchenogo Soveta instituta, TsGANTD SPb (Central State Archive of Scientific and Technical Documentation in Saint Petersburg), F. 303, Op. 13, D. 284, L. 44.

8. Aleksandr Delimov, *Drevesina i osnovnye produkty ee pererabotki* (Moscow: Goslesbumizdat, 1958), 43.

9. Proekt osnovnykh napravleniy razvitiya lesnoi, derevoobrabatyvayushchei i tsellyulozno-bumazhnoi promyshlennosti na 1976–1990 gg., RGAE (Russian State Archive of Economics), F. 4372, Op. 66, D. 5850, L. 6.

10. Otchet za 1978 god po teme "Podbor otechestvennykh taroupakovochnykh materialov v zamen importnykh," RGAE, F. 910, Op. 2, D. 372, L. 1b.

11. Narodnoe khozyaistvo SSSR za 70 let (Moscow: Finansy i statistika, 1987), 15.

12. Dzhon Steynbeck, *Russkiy dnevnik* (Moscow: E, 2017), 170.

13. V. Tatarinov, "Tekhnicheskiy progress—osnova razvitiya proizvodstva," *Lesnaya promyshlennost'*, no. 6 (June 1969): 1.

14. Doklad V. A. Valas'eva "O neotlozhnykh zadachakh v oblasti lesoekspluatatsii i plane nauchno-issledovatel'skikh rabot na 1962 god," RGAE, F. 9480, Op. 7, D. 532, L. 59.

15. Siegfried Giedion, *Mechanization Takes Command* (Oxford: Oxford University Press, 1948): 1.

16. Pervoe zasedanie sektsii sistem upravleniya tekhnologicheskimi protsessami v promyshlennosti na 1963 god, RGAE, F. 9480, Op. 7, D. 1226, L. 41.

17. Olessia Kirtchik, "From Pattern Recognition to Economic Disequilibrium: Emmanuil Braverman's Theory of Control of the Soviet Economy," *History of Political Economy*, no. 51 (2019): 186.

18. B. Ivanov, "Matematicheskie metody i vychislitel'nuyu tekhniku v obosnovanie razmeshcheniya lesnoi i lesopererabatyvayushchei promyshlennosti," *Lesnaya promyshlennost'*, no. 4 (April 1966): 4.

19. Dmitriy Teterin, *Problemy povysheniya effektivnosti primeneniya ASU v otrasli* (Moscow: VNIPIEIlesprom, 1979): 33.

20. A. B. Mnushkin, *Avtomatizatsiya proizvodstva i formirovanie novogo tipa rabotnika v SSSR. Avtoreferat dissertatsii na soiskanie uchenoi stepeni k.f.n.* (Moscow: Institut filosofii AN SSSR, 1961), 14–15.

21. Mnushkin, *Avtomatizatsiya proizvodstva i formirovanie novogo tipa rabotnika*, 7–8.

22. Yuliy Dukhon, *Avtomatizatsiya upravleniya v lesnoi promyshlennosti* (Moscow: Lesnaya promyshlennost', 1989), 3.

23. N. A. Medvedev, "Sistema-dokument-mashina," *Lesnaya promyshlennost'*, no. 6 (June 1973): 18.

24. N. A. Medvedev, "Nauka upravleniya," *Lesnaya promyshlennost'*, no. 2 (February 1973): 3.

25. Teterin, *Problemy povysheniya effektivnosti primeneniya ASU v otrasli*, 13.

26. V. Z. Gabriel, "Poisk vedet EVM," *Lesnaya promyshlennost'*, no. 5 (May 1981): 26.

27. "Za dal'neishuyu mekhanizatsiyu i avtomatizatsiyu proizvodstva," *Derevoobrabatyvayushchaya i lesokhimicheskaya promyshlennost'*, no. 7 (July 1953): 1.

28. Sokrat Punegov, *Razvitie tekhnicheskogo progressa na Kamennogorskoi bumazhnoi fabrike* (Leningrad: B. i., 1967), 22.

29. V. Zelenin, "Ubrat' pregrady s puti avtomatiki," *Lesnaya promyshlennost'*, no. 9 (September 1963): 4.

30. "Za vysokuyu kompleksnuyu proizvoditel'nost' truda," *Lesnaya promyshlennost'*, no. 9 (September 1953): 1.

31. "Modernizatsiya oborudovaniya—vazhnyi rezerv uvelicheniya vypuska produktsii," *Derevoobrabatyvayushchaya promyshlennost'*, no. 7 (July 1956): 2.

32. N. Serov, "Derevoobrabotka," *Derevoobrabatyvayushchaya promyshlennost'*, no. 5 (May 1978): 1.

33. Punegov, *Razvitie tekhnicheskogo progressa na Kamennogorskoi bumazhnoi fabrike*, 22 [emphasis added].

34. Zelenin, "Ubrat' pregrady s puti avtomatiki," 3–4.

35. N. A. Medvedev and A. S. Ditkovskiy, "Ispol'zovat' rezervy, vypolnit' plan," *Lesnaya promyshlennost'* 2 (February 1962): 1.

36. Protokol partiynogo sobraniya tresta "Sevzaples," 1955 god, TsGAIPD (Central State Archive of Historical and Political Documentation), F. 2951, Op. 1, D. 30, L. 26.

37. Protokol zasedaniya partkoma Svetogorskogo TsBK, 1964 god, TsGAIPD, F. O-1542, Op. 4, D. 154, L. 153.

38. Doklad "Partiyno-organizatsionnuyu rabotu—na uroven' novykh zadach!," 1957 god, TsGAIPD, F. O-1542, Op. 4, D. 18, L. 34.

39. Doklad L. V. Rossa, 1957 god, TsGANTD Samara (Central State Archive of Scientific and Technical Documentation in Samara), F. 342, Op. 1–1, D. 229, L. 116.

40. L. M. Maklykov, "Sokrashchenie ruchnogo truda—vazhnaya ekonomicheskaya i sotsial'naya zadacha," *Lesnaya promyshlennost'*, no. 8 (August 1983): 3.

41. Protokol mezhvedomstvennykh soveshchaniy v GKNT o razvitii lesopromyshlennogo kompleksa, 1984 god, RGAE, F. 9480, Op. 13, D. 1583, L. 187.

EPILOGUE

1. "Skandinavskaya lesnaya kontseptsiya v lesakh Rossii," *Lesnaya promyshlennost'*, no. 7 (1990): 2–3.

2. A. Orlov, "Severnyi les: defitsit pri izobilii," *Lesnaya promyshlennost'*, no. 1 (January 1990): 24.

3. Margarita Parfenenkova, "60 ekologicheskikh organizatsiy zastupilis' za Vsemirnyi fond prirody," Vedomosti. Ustoichivoe razvitie, accessed August 08, 2023, https://www.vedomosti.ru/esg/protection_nature/articles/2023/03/21/967422-bolee-60-ekologicheskih-organizatsii-zastupilis-vsemirnii-fond-prirodi.

4. F. T. Chomaev, "Razvitie lesopromyshlennogo kompleksa Rossiyskoi Federatsii na fone obshchemirovykh tendentsiy," *Vestnik GUU*, no. 10 (2017): 47–51.

5. I learned about this in private talks during my visit to Svetogorsk in 2013 and 2019.

6. V. Spivak, "Velikaya kitaiskaya vyrubka," Carnegie Endowment for International Peace, accessed February 12, 2023, https://carnegie.ru/commentary/77100.

7. A. Gruber, "Lesozagotovka est'—pererabotki otkhodov net," accessed April 9, 2018, https://lpk-sibiri.ru/bioenergetics/pellet-plants/lesozagotovka-est-pererabotki-othodov-net.

8. See, for example, "Borshchevik spaset: uchenye nashli primenenie sornyaku na fone sanktsiy," Infox, accessed February 12, 2023, https://infox-ru.turbopages.org/infox.ru/s/news/283/271597-borsevik-spaset-ucenye-nasli-primenenie-sornaku-na-fone-sankcij.

9. *Obzory po informatsionnomu obespecheniyu tselevykh kompleksnykh nauchno-tekhnicheskikh program i program po resheniyu nauchno-tekhnicheskikh problem* (Moscow: NII problem vysshey shkoly, 1991): 1.

10. Lars Carlsson and Mats-Olov Olsson, "Initial Analyses of the Institutional Framework of the Russian Forest Sector," *Interim Reports*, IR-98-027 (1998): 8.

11. V. Glotov, "Vozrozhdenie neperspektivnykh lesnykh poselkov," *Lesnaya promyshlennost'*, no. 2 (February 1990): 8.

12. Peder Anker, *The Power of the Periphery: How Norway Became an Environmental Pioneer for the World* (Cambridge: Cambridge University Press, 2020).

13. Frank Oekötter, *The Greenest Nation? A New History of German Environmentalism* (Cambridge, MA: MIT Press, 2014).

14. Arvid Nelson, *Cold War Ecology: Forests, Farms, and People in the East German Landscape, 1945–1989* (New Haven, CT: Yale University Press, 2005).

15. John Bellamy Foster, *The Return of Nature: Socialism and Ecology* (New York: Monthly Review Press, 2021), 211.

16. See, for example, Donald Worster, *Dust Bowl: The Southern Plains in the 1930s* (Oxford: Oxford University Press, 1979).

17. Zsuzsa Gille, *From the Cult of Waste to the Trash Heap of History: The Politics of Waste in Socialist and Post-Socialist Hungary* (Bloomington: Indiana University Press, 2007).

18. Stephen Brain, "Stalin's Environmentalism," *Russian Review* 69, no. 1 (Winter 2010): 93–118.

19. Petr Jehlička and Joe Smith, "Out of the Woods and into the Lab: Exploring the Strange Marriage of American Woodcraft and Soviet Ecology in Czech Environmentalism," *Environment and History* 13, no. 2 (2007): 187–210.

INDEX

Note: Page numbers in italic indicate figures.

Abundance
  cultural myth versus economic reality, 29–38, 55–58, 165–166
  imperial poetics of, xiii–xiv, xxxv–xxxvi, 1–2, 5, 29–30
Academy of Sciences, 30, 40, 83, 91, 93
Administrative change, 21–27
Agricultural by-products, 70–71
Alarmism
  as challenge to resourceful imperialism, xxix, xxxv, 77–78, 165–166
  discourse of, xx, xxvi–xxvii, 165–166
  factors contributing to, 5–11, 139
All-Union Project Institute of Forestry, 42
Altai forests, 33, 64
Alternative raw resources. *See* Experimental solution; Reed and annual plants; Waste materials
Aluminum, 47, 79
Amazon rain forest, xx–xxi
Amur region, 30, 47
Amursk pulp plant, 47–48
Angara-Yenisey (river system), 46, 49
Angara River, 46, 49
Angara-Yenisey territorial-industrial complex, 46–47
Anker, Peder, 170
Annual plants. *See* Reed and annual plants
Anuchin, Nikolay, 97
Ash, 67
Asian regions, 7, 57, 109
Aspen, 65, 108
Assembly lines, 131, 140–141
Asteryakov, G., 116
Astrakhan'
  brickmaking factories in, 111
  reed-based enterprises in, 104, 116–117, 122–129

Authoritarian regimes, nature-economy relations of, xxiii–xxiv
Automation
  conflict between professional visions and implementation, 148–155, 168–170
  forestry sector implications of, xxxii–xxxiii, 145–149
  industrial applications of, 139–148
  rationality as ideological basis of, 148, 166–169
  Soviet-Western comparisons, 11–12, 140–142
  technical infrastructure for, 139–140
Automobile transportation, roadbuilding for, 38–45

Baikal (lake), 20, 32, 47–48, 99–101
Baikal-Amur Mainline (BAM), xxix–xxx, 39–40, 47
Baikal'sk pulp and papermaking plant, 32, 47–48, 99–101
Balakhtash, 60
Baltic region, 47, 105
Bark, 11, 39, 82, 89
Barr, Brenton, xxiii
Basic Law on Forests (1923), 61
Beregometsk forest combine, 90
Biofuel, waste materials used in, 82–83
Biryusa River, 49
Bisert' logging company, 152
Bogdanov, V., 126

Bolsheviks, 137, 140
Borisova, A. P., 128
Bowles, Donald, 11–12
Braden, Kathleen, xxiii
Brain, Stephen, 170
Branches, waste from, 11, 92, 136
Bratsk hydropower station, 49
Bratsk LPK, 47, 52, 60–61, 101
Brezhnev, Leonid, 53, 64
Brik, M. I., 43
Bulgaria, 52
Buryatia region, 62, 66

Cabinetry, waste materials used in, 80–81
Calculating machines, 142–144
Capitalism
  modernization in, 136–137, 140–141, 145–146
  nature-economy relations in, xvii, 2, 170–171
  Soviet-Western comparisons, 136–137, 140–142, 149, 154
  waste in, 71, 75
Carbon raw materials, xxix
Cardboard
  consumption of, 8, 84–85, 138
  industrial manufacture of, xxxiii, 16, 47–48, 76, 79–80, 104, 106, 164
Carson, Rachel, 95
Cedar, 33
Celluloid, 134, *135*
Cellulose-making industry
  administrative change in, 22–23

INDEX

alternative raw resources for, xxviii, 106, 163
consumer products, xxxiii, *81*
importance of, xv, xvii–xviii
industrial capacities, 47–48, 56, 105, 127
production of, 132
raw material shortages for, 41
Central Committee, 40, 98, 126
Central Institute of Cellulose and Paper Industry, 84
Central Research Institute of Reed, 125
Chain saws, 12
Chemistry and chemicalization
modernization with, xxxiii–xxxiv, 132–137
in no-waste technological production, 78–80
in post-Soviet era, 162–163
Chepetsk logging company, 151
Chernobyl catastrophe, xxiii, 101
China
alternative raw resources in, 111
wood exports to, 161
Chipboard, waste materials used in, 80
Chistyakov, Nikolay, 108
Cities and city-like settlements, 53, 55, 59–60
Clean-cuttings, 25–27
Clear-cutting
factors contributing to, 25–26
in post-Soviet era, 157, 161–162, *162*
in Siberia and Far East, 59, 60, 62–63
sustainability and waste, 92, 169
Clothing, forestry worker, 15
Cold War
alternative raw materials and, 105–106
economic importance of wood during, xv–xvii, 4–11, 13, 79
waste culture and, 86–87
Collective farms, forest resources of, 23–24
Colonization. *See* Imperial solution
Committee on Forestry and Wood Processing Industry, 40
Communist Party, 50, 73, 98, 144
Complex use of natural resources, xxxiii, 69–77, 163–165
Coniferous wood, 63–65, 75
Consumerism, shift toward, xxxiii, 35, 76, 79, 99, 132, 164
Consumerist vision of nature, xviii, 4–5, *5*, 95, 98, 106
Consumer waste, 83–89. *See also* No-waste technological production
Copper, 79
Council of Ministers, 15, 40, 98, 126
Council on Exploration of Productive Forces of the Academy of Sciences, 50

Council on the Study of Productive Forces, 30
Cow parsnip, cellulose production from, 163
Cybernetics, 143, 147, 154
Czarism, 35
Czech Republic, 171

Dal'lesprom industrial merger, 99
Danube River, 109, 112, 117, 119
Deciduous wood, 75–76
Deforestation
  consequences of, 17, 59–60, 73, 78, 101, 105–106
  emotional responses to, xxvi–xxvii, xxvii
  global rate of, xx–xxi, 22
  in northwest region, 6–7, 20–21, 25–26
  in post-Soviet era, 158–161
  in Siberia and Far East, 59
Denmark, xxi
De-Stalinization program, 14, 73
Dnieper River, 109, 119
Dniester, 109
Don River, 109

East Germany, 52, 110–111
Ecocidal view of socialist forestry, xxiii, 170
Ecological activism, 101
Ecology. *See* Industrially embedded ecology
Economic Automatic Management System, 142

Economic sanctions on Russia, 162
Economizing, discourse of, 51, 69–75, 78, 97, 167
Egypt, xvii
"Electric paper," xiv–xv
Electrification, in Siberia, 46–47
Electronic calculating machines, 142–144
Energy needs, Siberia, 46–47
Expansion of forestry industry
  conflict between professional visions and implementation, 55–67, 168–170
  forestry industry enterprises, 46–55, *54*
  natural resources and image of green abundance, xiii–xiv, 7–8, 29–38, *54*, 109, 165–166
  overview of, xxix–xxx
  rationality as ideological basis of, 36–37, 50, 55–57, 166–169
  social and urban infrastructure, 53, 55, 59–60
  as tabula rasa to build new enterprises, 29–38
  transportation infrastructure, 38–45
Experimental solution. *See also* Alternative raw resources; No-waste technological production; Reed and annual plants
  conflict between professional visions and implementation, 91–94, 121–129, 168–170

discourse of economizing and, 69–75, 97, 167
industrially embedded ecology in, 167–168
overview of, xxx–xxxii
rationality as ideological basis of, 69–77, 166–169
Exports, timber
in post-Soviet era, 157–159, 161–162
sources of, 9, 48
volume of, xx, xxxvi–xxxvii, 16, 94, 139
Extractive economy, xvi–xvii, xxii, xxv, 2, 50

Far East (Soviet Union), expansion of forestry industry to
conflict between professional visions and implementation, 55–67, 168–170
forestry industry enterprises, 46–55, *54*
industrially embedded ecology in, 167–168
natural resources and image of green abundance, xiii–xiv, 7–8, 29–38, *54*, 109, 165–166
overview of, xxix–xxx
rationality as ideological basis of, 36–37, 50, 55–57, 166–169
social and urban infrastructure, 53, 55, 59–60
as tabula rasa to build new enterprises, 29–38
transportation infrastructure, 38–45
waste in, 92
Fertilizer, waste materials used in, 82
Fiberboards, 80–81, 93, 106, 126
Finland, 2
enterprises annexed from, 47, 105
road construction in, 42
wood harvesting in, 13, 159
Fir, 7, 9, 31–33, 46, 64, 101, 157
Fishery nets, papermaking from, 104–105
Five-year plans, 3, 41, 46, 116
Flooding of reedbeds, 122
Food industry, wastepaper use in, 86–87
Fordism, 140
Forest Code (1923), 61
Forest Code (1978), 61
Forest-industrial complexes. *See* LPKs (forest-industrial complexes)
Forestry Industry, 128–129
Forestry labor, 14–15
Forestry management system, administrative change in, 21–27
Forests. *See also* Alarmism
categories of, 2–3
dying/dead metaphor for, 25–26
eastern forest reserves, xiii–xiv, 7–8, 29–38, *54*, 109, 165–166
economic role of, xiii–xix

Forests (cont.)
  geographic distribution of, 1–8, *66*
  geography and distribution of, *54*
  image of abundance of (*see* Abundance)
  industrialist approach to, 1–11
  liminality of, xxxiv, 94–101
  productivist approach to, xxvi, 74, 101
  superpowerness of, 164–165
France, 53, 55
Franco, Francisco, xvi
Fuel, wood used for, xv
Furniture making, 22–23, 80–81, 162–163

Gabriel, V. Z., 146
Geography of forests, 1–6
German Democratic Republic, xxiii, 13, 55, 104
Giedion, Siegfried, 141
Gigantomania, 22–23, 48
Gille, Zsuzsa, 170
Giprobum, 89, 108–109, 124
Glushkov, Viktor, 142
Gorbachev, Mikhail, xvii
Gorbachev, V., 59
Gorbatov, E. I., 66
Gosplan, 93–94

Harlu, 23
"Holiday of the Forest, The," (Marshak), 3–4
House construction, xv–xvi, 80–81, 93

Human factor, in "wood crisis," 18–21
Hungary, 52, 170
Hydropower, 46–47

Ilim River, 49
Imperial solution. *See* Expansion of forestry industry
Industrialist approach
  economizing imperative in, 51, 69–75, 97, 167
  nature-economy relations in, 1–11, 29–32, 76–77
  sustainable forestry discourse in, 95–101
Industrial legislation, 61, 98–100
Industrially embedded ecology. *See also* Alarmism; Alternative raw resources; Modernization; No-waste technological production; Wood crisis
  concept of, xxv
  industrial dimension of, xix–xxix
  in post-Soviet era, xxxvi–xxxvii, 159–160, 169–170
  Soviet path to, xiii–xix, xxxiv–xxxv, 26, 159–160, 167–170
Industrial waste. *See* No-waste technological production; Waste materials
Institute of Forest and Wood, 40
Institute of the Paper-and Cellulose-Making Industry, 110, 114

Institute to Project Papermaking Enterprises, 40
Institutional reorganization, state-led, 21–27
Irkutsk region, 2, 39, 50, 52–53
Irtysh River, 109
Ismail pulp plant, 114, 117, 125
Italy, 111

Japan
  enterprises annexed from, 47, 105
  forestry equipment imported from, 13
  Toyota Corporation, 52

Kakhovsk water reserve, 118
Kalganak, 59–60
Kamennogorsk papermaking factory, 150
Kanevskiy, M., 56
Karelia
  administrative change in, 23
  waste in, 93
  wood harvesting in, 20–21, 25–26, 48, 62, 157
Karymsakov, V., 125
Kazakhstan, xxxi, 37, 56, 103, 106, 117
Kharkevich, Aleksandr, 142
Kherson pulp plant, 108, 114, 117, 122
Khrushchev, Nikita, 21, 87, 109, 116
Kitov, Anatolyi, 142
Kola Peninsula, 62
Kolkhozy, 23–24
Kolosovskiy, Nikolay, 49
Komsomol Youth organization, 53
Kondopoga cellulose and papermaking plant, 47
Kostroma plywood plant, 91
Kosygin, Alexey, 116
Kosygin economic reform, 143
Kovalev, N. I., 142
Kovaleva, D., 88
Krasnoyarsk region, 60, 93
Kremenchugsk water reserves, 118
Kuban', 109
Kubenskiy, A., 100
Kuteinikov, Feodor, 84–86, 114
Kuzmishchev, G., 111
Kyucharyants, V., 98
Kzyl-Ordynsk pulp and cardboard-making plant, 117, 125

Labor
  availability and working conditions, 14–16
  modernization and, 149–152
  seasonal workers, 14–15, 42–43, 114–115, 150
  in Siberia and Far East, 47, 52–53
Larch, 65
Latour, Bruno, 174n12
Lean manufacturing, 52
Legislation, industrial, 61, 98–100
Lena River, 49
Lenin, Vladimir, 31

Leningrad Research Institute of Urban Development, 53
Leonov, Leonid, 10
Lespromkhozy, 20, 22, 48, 66, 119
Liminality of forests, xxxiv, 94–101
LPKs (forest-industrial complexes), 50–55
Lyaskelya papermaking factory, 23

"Main Directions of Economic and Social Development of the USSR," 98
Maklakovo-Yeniseysk region, 91–92
Malyshev, V. M., 46
Marshak, Samuel, 3–4
Marx, Karl, 144
Marxism, 100, 144–145
Mechanization
  conflict between professional visions and implementation, 148–155, 168–170
  industrial applications of, 139–148
  overview of, xxx–xxxii
  rationality as ideological basis of, 148, 166–169
  Soviet-Western comparisons, 140–142, 149
  technical infrastructure for, 11–14
  workers' rationalization movement and, 69
Mechanization Takes Command (Giedion), 141

Medvedev, N., 146
Militarism, 164
Military applications of wood, xv, xviii–xx, 23, 79
Mineral resources, 31
Ministry of Food Industry, 86
Ministry of Forestry, 21–27, 43, 74, 81
Ministry of Transport, 23
Miynala, 157
Mnushkin, A., 144
Modernization
  automation (*see* Automation)
  conflict between professional visions and implementation, 148–155, 168–170
  ideology of, 131–139
  industrially embedded ecology in, 167–168
  mechanization (*see* Mechanization)
  overview of, xv–xix, xxxii–xxxv
  perestroika, xvii, 20, 158–160, 169–170
  rationality as ideological basis of, 132–133, 148, 166–169
  Soviet-Western comparisons, 136–137, 149, 154
Mudrik, Viktor, 86, 89, 106

National Automated System for Computation and Information Processing (OGAS), 142–143, 147
Netherlands, 111
Newspaper, 86–87

Northwestern forests
  deforestation in, 6–7
  deforestation of, 20–21, 25–26
Norway, 2, 164, 170
Novosibirsk region, 109
No-waste technological production, 77–83
  chemistry and chemicalization in, 78–80, 134–136
  conflict between professional visions and implementation, 91–94, 168–170
  consumer waste and recycling, 83–89
  discourse of economizing in, 69–75, 78, 97, 167
  environmentalist attitudes arising from, 94–101, 167–168
  industrial legislation related to, 98–100
  productivist approach to, xxvi, 51–52, 74, 101
  rationality as ideological basis of, 69–77, 163–165, 166–169
  sustainable forestry, 95–101
  waste culture, 85–88, 100–101
  wood and sawmill waste, 89–94

Ob' River, 109
Oekötter, Frank, 170
Omsk region, 109
Orlov, Georgiy, 83, 116

Packaging, 76, 106, 138
Paper and Wood Processing Industry, 83
Papermaking
  administrative change in, 22–23
  chemicalization of, 131–137
  consumer paper and paper consumption, 84, 86, 124, 137–138, 162–164
  "electric paper," xiv–xv
  industrial capacities, xxxiii–xxxiv, 9–10, 47
  insufficient manufacture of, 137–139
  mechanization and automation in, 150
  overview of, xvii–xviii
  in post-Soviet era, 161–162
  reed and straw used in, 104–106, 124
  in Siberia and Far East, 47–48
  waste materials used in, 79–80
Paper waste
  industrial value and recycling, 83–89
  volume of, 84–85, 89
Perestroika, xvii, 20, 158–160, 169–170
Pine, 64, 82
Plakhov, I., 153
Planned economy
  industrialism in, 1–11
  modernization in, 142–144, 149–150
  overview of, xxxvi–xxxvii

Planned economy (cont.)
post-Soviet era compared to, 157–160
waste in, 74–75
Plastics, wood, 79–80, 134, *135*
Plywood, 4, 91, 93, 138
Poland, 52
Political regimes, nature-economy relations and, xxiii–xxiv, 170–172
Pollution
Baikal'sk pulp and papermaking plant, xxiii, 32, 47–48, 99–101
clear-cutting and, 62–63
environmentalism in response to, 20, 99–101
in post-Soviet era, 159, 161
Post-Soviet era
forestry industry in, 157–162, *162*
industrial "ecologization" in, xxxvi–xxxvii, 160–163
Soviet planned economy compared to, xxxvi–xxxvii, 157–160
Prison labor, 14
Productivist approach, xxvi, 74, 101
Professional dreamscapes, conflict with infrastructural realities
expansion of forestry industry, 55–67
lessons and insights from, 163–166
modernization, 148–155
no-waste technological production, 91–94
overview of, xiii–xix
reed-based enterprises, 121–129
Progressivists, 152, 154, 155
Project and Technological Institute of Furniture, 81
Protected forests, 2–3
Pulp-based products
alternative raw resources in, 107–108
insufficient manufacture of, 137–139
in post-Soviet era, 162–163
postwar growth of, 9
Siberian and Far East enterprises, 47–48
Punegov, Sokrat, 150, 151
Putikov, M. A., 104
Putin, Vladimir, 160

Railroad construction, xxix, 25, 38–46, 49, 66
Rationality
concept of, xxi, xxx
eastern expansion and, 36–37, 50, 55–57, 165–166
environmentalism and, 99–101
lessons and insights from, 163–165
modernization and, 132–133, 148, 166–169
no-waste technological production and, 69–77
reed-based enterprises and, 103, 107, 115–116, 124, 129

Rationalization movement, 69
Rational system of economic control, 142
Recycling, 85–88, 108
Reed and annual plants
  conflict between professional visions and implementation, 121–129, 168–170
  cost of, 104
  cow parsnip, 163
  distribution and abundance of, 103, 106–110
  harvesting of, 112–116, *113*, 119–121, *120*
  industrially embedded ecology in, 167–168
  industrial promise of, 103–108
  overview of, xxx–xxxii
  productivity and shortages of, 121–129
  as rational resource, 103, 107, 115–116, 124, 129, 166–169
  reed-based enterprises, construction of, 116–121
  research and experimentation with, 108–116, *113*
  straw, 70, 103–104, 110–111
  sustainability of, 104, 110, 121, 128–129
Reforestation
  factors hindering, 44, 62–63, 154
  lack of, xxxvi, 25–26, 67
  in post-Soviet era, 161–162
  productivist approach to, 74, 96

Reorganization of forestry management system, 21–27
Research Institute of Forestry, 46
Resource abundance. *See* Abundance
Resourceful imperialism, 165–166
Roadbuilding, 38–45
Romania, 52, 111–112
Ross, L., 154
Rubber, xv, 85–86, 136
Russian Forest, The (Leonov), 10
Russian Journal, A (Steinbeck), 139

Safety gear, forestry worker, 15
Sanitary towels, 138
Sawmill processing. *See also* No-waste technological production
  administrative change in, 22–23
  waste in, 89–94
Saws, 12
Scandinavian type of cutting, 159
Scientific-Technical Council of Paper Industry, 70
Scientific-Technical Society of the Forestry Industry, 143
Seasonal workers, 14–15, 42–43, 114–115, 150
Segezha cellulose and papermaking plant, 47
Selenga River, 64
Selenginsk pulp and cardboard plant, 47

Settlements, 53, 55, 59–60
Sevzaples, 153
Shcherbakov, A., 96–97
Shcherbina, G. I., 87
Shipbuilding, xv, 3
Shiryaev, V., 19–20
Shock-work campaign, 47–48, 52–53
Siberia, expansion of forestry industry to
 conflict between professional visions and implementation, xxvii, 55–67, 168–170
 forestry industry enterprises, 46–55, *54*
 industrialist approach to, 29–32
 natural resources and image of green abundance, xiii–xiv, 1–8, 29–38, *54*, 109, 165–166
 overview of, xxix–xxx
 in post-Soviet era, 161
 rationality as ideological basis of, 36–37, 50, 55–57, 166–169
 social and urban infrastructure, 53, 55, 59–60
 as tabula rasa to build new enterprises, 29–38
 transportation infrastructure, 38–45
 waste in, 91–92
Silent Spring (Carson), 95
Silk, synthetic, 82, 133
Sinyaev, N. V., 93
Smith, Jenny, 125
Social infrastructure, 53, 55, 59–60

Socialism, changing view of nature-economy relations in, 169–172. *See also* Industrially embedded ecology
South (Soviet Union), reed-based enterprises in, xxx–xxxii, 103–104, 106, 110, 116–121
Soviet-Finnish border enterprises, 13–14
Soviet-Finnish War, 23, 157
Soviet planned economy. *See* Planned economy
Sovkhozy, 23–24
Sovnarkhoz reform, 21–23
Spain, xvii
Specialists
 alarmism among (*see* Alarmism)
 definition of, xvii
 dreamscapes of (*see* Professional dreamscapes, conflict with infrastructural realities)
 responses to wood scarcity (*see* Wood crisis)
Specialist solutions. *See* Experimental solution; Imperial solution; Modernization
Stalinist industrialization, xxx–xxxii, 48–49, 170
State-led institutional reorganization, 21–27
Steinbeck, John, 139
"Storm over Baikal." *See* Baikal'sk pulp and papermaking plant
Straw, 70, 103–104, 110–111

Sukachev, Vladimir, 40, 84
Sukhona River, 64
Sunken logs, recovery of, 127–128
Superpowerness of forests, 164–165
Surgut, 46
Sustainability, xxxiii
  of forestry industry, 20–21, 60–62, 95–101
  of reed, 104, 110, 121, 128–129
Sustainable forestry, 20–21, 95–101
Sverdlovsk region
  Bisert' logging company, 152
  Svetogorsk pulp and papermaking plant, 153, 161
Sweden, xxxvi, 2, 7, 95, 111
Synthetic polymers, 48–50, 133–134

Tabur River, 62
Taiga. *See* Forests
Tatarinov, V. P., 41
Taylorism, 140
Technical improvement. *See* Modernization
Technical Management of the Forestry Ministry, 154
Territorial administration reform, 21–22
Territorial-industrial complexes (TPK), 48–50
Teterin, D. I., 143, 146
Tikhomirov, B., 45
Timber exports. *See* Exports, timber
Timofeev, Nikolay, 29
Totalitarian regimes, nature-economy relations of, xxiii–xxiv
Toyota Corporation, 52
Tractors, 12
Transportation infrastructure, 38–45
Tree bark, 11, 39, 82, 89
Tree species, distribution of, 66. *See also individual species*
Tsyurupinsk
  agricultural and biological stations in, 119
  Kherson pulp plant in, 108, 114, 117, 122
Tyumen', 46

Ukraine
  Beregometsk forest combine, 90–91
  Ismail plant, 125
  Kherson pulp plant, 108, 114, 117, 122
  reed-based enterprises, xxxi, 108–109, 117, 122, 125
  reedbeds in, 103, 106, 109
  Russian war in, 160
  transportation of materials to, 56
Unified Communication System, 142
United States
  consumer society in, xv, 133
  economic sanctions on Russia, 162–163

United States (cont.)
  mechanization and automation in, 11–14, 140–142
  paper consumption in, 137–138
  papermaking in, 16
  recycling and waste culture, 86–87
Upper Pechora railroad, 40
Ural region, 7–8
Ural'sk logging company, 150
Ust'-Balyk-Omsk industrial complex, 47, 52–53
Ust'-Balyk-Omsk oil pipeline, 46
Ust'-Ilimsk project, 53, 55, 60–61

Varaksin, Feodor, 84, 108
Vibration disease, 12
Viscose, 31, 134
Vlasov, G., 101
Volga hydropower station, 121
Volga River, 108–111, 118, 119
Volkov, O. V., 64

Waste culture, 86–87
Waste materials. *See also* No-waste technological production
  consumer waste and recycling, 70, 83–89
  environmentalist attitudes arising from, 94–101
  industrial applications of, 77–83
  industrial legislation related to, 98–100
  as material of modernity, 77–83
  rational and complex use of, 69–77, 163–165
  in Siberia and Far East, 56
  volume of, 11–12, 84–85, 89
  waste culture, 100–101
  wood and sawmill waste, 17, 89–94
Wastepaper
  industrial value of, 70, 83–89
  recycling of, 85–88
  volume of, 84–85, 89
What Chemistry Makes from Wood, 134
Williams, Rosalind, xxv
Wittenberg, straw cooking in, 110–111
Women, in forestry labor, 14–15
Wood-based products
  consumer demand for, 161–163
  insufficient manufacture of, 15–19, 137–139
  postwar growth of, 8–9
Wood crisis
  alarmism about (*see* Alarmism)
  factors contributing to, xxv–xxvi, xxxiv–xxxv, 9, 11–21, 26–27
  in post-Soviet era, 157–163
  productivist approach, 74, 101
  solutions to (*see* Experimental solution; Imperial solution; Modernization)

Wood-harvesting industry, xxxii
    administrative change in, 21–27
    distance from wood-processing operations, 23–25
    labor for, 14–16, 149–152
    mechanization and automation in, 150–151
    modernization of (*see* Modernization)
    in post-Soviet era, 157–160, *162*
    productivity issues in, 11–21
    in Siberia and Far East (*see* Expansion of forestry industry)
    waste in (*see* No-waste technological production)
Wood plastics, 79–80, 134, *135*
Wood-processing industry
    administrative change in, 21–27
    economic role of, xiii–xix
    industrial dimension of, xix–xxix
    industrialism and, 1–11
    military significance of, xv, xviii–xx, 79
    modernization of (*see* Modernization)
    in post-Soviet era, 161–162
    waste in (*see* No-waste technological production)
"Wood to Construction Sites" (Kyucharyants), 98
Wood waste, 17
    industrial applications of, 90–94
    volume of, 89–91

Wood-wool, 80
Workforce, 14–15
    availability of, 149–152
    in Siberia and Far East, 47, 52–53
Work stoppages, 15

Yakutia, 4
Yakut people, 29–30
Yenisey, 109
Yugoslavia, 2

Zelenin, V., 152